W9-CJM-425

A Dog Named Beautiful

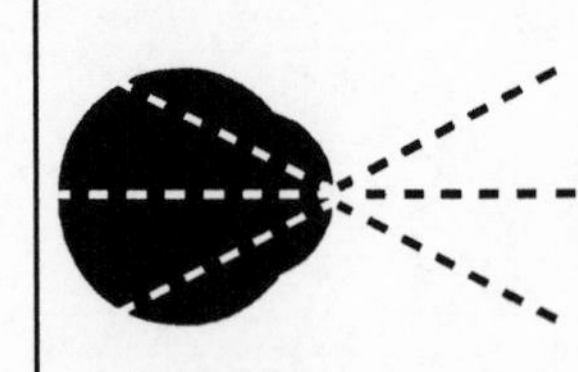

This Large Print Book carries the Seal of Approval of N.A.V.H.

A Dog Named Beautiful

A Marine, a Dog, and a Long Road Trip Home

Rob Kugler

THORNDIKE PRESS
A part of Gale, a Cengage Company

Farmington Hills, Mich • San Francisco • New York • Waterville, Maine
Meriden, Conn • Mason, Ohio • Chicago

This is a true story. Some dates and exact locations or events depicted in the book have been changed.

Thorndike Press® Large Print Bill's Bookshelf.
The text of this Large Print edition is unabridged.
Other aspects of the book may vary from the original edition.
Set in 16 pt. Plantin.

**LIBRARY OF CONGRESS CIP DATA ON FILE.
CATALOGUING IN PUBLICATION FOR THIS BOOK
IS AVAILABLE FROM THE LIBRARY OF CONGRESS**

ISBN-13: 978-1-4328-6765-2 (hardcover alk. paper)

Published in 2019 by arrangement with Macmillan Publishing Group, LLC/Flatiron Books

Printed in Mexico
1 2 3 4 5 6 7 23 22 21 20 19

To all those who dream big and love hard, who have experienced the emptiness of life after loss and a world without dreams, who refuse to live in that reality, and strive to create a new one

CONTENTS

PART THREE: UNCONDITIONAL LOVE

pinned. My defenses are useless. There's no going back to sleep after this.

"All right, all right," I say. "I'm up, I'm up." I sit up and stretch my arms.

Thud, thud, thud. Her tail whips against the window. "Yeah, Daddy, yeah! That's it! Keep going!"

I look at her again, this time more closely. My tired and weary face is quickly replaced with a happy and hopeful one. I let out another big sigh as I wrap her up in my arms, squeeze her tight, and murmur, "I'm so lucky to have you, baby."

She squirms out of my arms and begins to bounce as she gives me a holler: "Cool, now feed me breakfast!"

"Ha ha. Yeah, yeah. I'm just the food guy."

I work my way over to her bucket of food and pour her a bowl of kibble. I slide on pants and a T-shirt while Bella gobbles her food, her tail wagging the entire time. Once she's finished, I put on my jacket and boots, open the tailgate, and help Bella down. A nearby trail in the woods beckons, and we head toward it to take a hike. The woods are silent, breathless in the early-morning air, and I snap off her leash and let her run free. Bella darts about from scent to scent, all the while wearing her patented smile, the smile that's stolen my heart.

Wow. How lucky I truly am, I think. This is how every morning on the road starts with this girl. I can't remember a time on this trip that I haven't been woken up by Bella's kisses. Whether it's because she wants to let me know she loves me, or because she merely needs me awake to appease her insatiable appetite, which runs on a strict internal schedule, she's always excited about mornings. She's simply excited for every moment she's alive.

We return to the 4Runner, which I've affectionately dubbed "Ruthie," set for another day of traveling. I open the side door and help Bella up into the passenger seat. I walk around the front of the vehicle, open the other door, and slide behind the wheel. The engine starts up confidently, and as we pull out of the parking lot, I notice that Bella is smiling. So am I. It's hard to have a bad day when each starts as perfectly as this.

I don't know for sure how long Bella has left, and, yes, this thought concerns me deeply. But all is well with Bella's health this bright morning, and I brush anxious thoughts from my mind. The sky is blue, the road is sure, and I remind myself that Bella and I are on a quest. Our journey is far from completed, and we need to keep going. I love this dog so much and want to take her on one last adventure so I can give her everything I possibly can in life. Yet I realize Bella is helping me too. She's helping me find something I so desperately need.

The road home.

▪ ▪ ▪ ▪

One: One Man and His Dog

▪ ▪ ▪ ▪

1
A New Life on Three Legs

"Can I talk to you in the back room about these X-rays?" the vet asks, and my optimism suddenly does a one-eighty. Nothing good ever needed to be talked about in a back room. We've come here for what I thought was a sports injury, and up until now, the prognosis has been nothing but positive.

I leave the waiting room and follow the vet down the hallway to the back of the clinic. Two orange–cinnamon spice teas and a hot chocolate slosh around in my gut. Whenever there are free drinks around, I can't help but take advantage, and it's been a long afternoon already in the vet's office. Bella walks beside me on a leash, her paws clicking on the linoleum. Bella is happy anywhere, as long as she's by my side. She has been my copilot on so many adventures together.

We reach the back room and I kneel to

pet Bella's silky-smooth fur, starting at the top of her head just above her eyes. I'm the type of person who treats his dog like family. I am not Bella's owner; I am her guardian and companion. My philosophy has always been that I'll do all I can to make sure she has a good life on this earth. Bella is always happy, always positive, always smiling, always up for adventure. We need nothing other than food, shelter, and each other's company. Her ears run between my index finger and thumb, and I play with the tiny pocket in the ear on the anterior side, near where it connects to her head. Ever since she was a puppy, I've loved to play with that little spot.

The vet flips on the lights to the X-ray panels, and I look up to see two pictures: Bella's humerus and Bella's lungs. On top of the humerus, the shorter bone in the forelimb, sit tiny spiderweb feathers. The bone looks to be growing tiny pieces of itself into the surrounding tissue, and Bella's lungs appear cloudy with bright white spots, like a marbled sky bracing itself against an approaching storm.

"I'm sorry," the vet says. "It's advanced osteosarcoma. It's the worst news I could give you. Bone cancer. It's spread to her lungs."

The words take the breath out of me like a punch to the stomach. I choke up but clench my jaw and hold my composure. Not because I'm too tough to cry. Not because I'm trying to portray some sort of hyper-masculine military persona. I hold my composure because I need to focus. I need to keep my emotions in check and listen to every word the vet says so I know exactly what next steps to take. Over the years I've learned that when the shit hits the fan, the world stops and I pay attention. If only I had this much clarity from day to day.

The thought of Bella being gravely sick is almost too much to take in. I look down and see this happy dog, her bright eyes and full smile. Bella looks back up at me. She is adventurous, athletic, an endearing ball of doggy love. She is my one constant. She loves me unconditionally when I am broken, when I am sad, when I am angry, when I am wrong, when I have failed. Her tail wags whenever I walk in the door, and over the past few rocky years of my life she has become a wise mentor to me, a Yoda to Luke, teaching me how little is needed in life to be happy, teaching me to be ever present in each moment. This dog has become my best friend, and now I'm going to lose her.

The vet's brow is furrowed. I look back into her eyes. I understand this must be the worst part of her job. "What are our options?" I ask, feeling my way forward.

"The cancer is aggressive. We can take the front leg. But if you don't want to take the leg, then we should put her down soon, because she's in so much pain."

Put her down? There's no way I'm going to put her down. Bella means the world to me. She's stuck with me through everything. She models the components of what it means to be alive — happiness, freedom, service, purpose, pleasure, joy. She prizes the very act of *being.* No, as long as there are viable options, I will not take this dog's life. But I can't bear the thought of Bella being in pain either.

"If we take the leg, how much time will she have?" I ask.

"The cancer will still be in her lungs. The surgery will only take her pain away. So even with taking her leg, you're looking at three to six months left at most."

Three to six months.

The news rushes at me like a bullet. Bella is only eight years old. It's not uncommon for a Labrador to live to twelve, or more. I glance at Bella again and pet her head. Her tail continues to wag. She's looking up at the vet, curious about what's causing the gloomy mood. Amputating her front leg is our only option, but that option seems so severe, and at the same time not enough. I ask about chemotherapy but am told that with the advancement in her lungs, chemo is pointless. I know I need to make a decision about the amputation right away, but it's killing me to say yes. I need time. Time to think. Time to weigh the pros and cons. I ask if I can consider the decision overnight.

Of course. Overnight. Bella and I drive home, and I feed her supper, then cook my own on a little hot plate. I call my family and friends for advice.

"That's too much money," says a friend. "I'd just put her down."

"What kind of quality of life will she have with only three legs?" says another. "Put her down."

"Put her down," says a third.

"Put her down."

"Put her down."

"Put her down."

But I can see the life left in this dog's eyes. No. I won't put her down. She's nowhere near ready. I look up information online and see how great results can come from this procedure, although I read a few stories of a short life afterward as well. "We can do this, Bella girl. I've got you," I whisper to her.

My mind works overtime. It seems like only yesterday we got Bella. An ad in a local newspaper caught my girlfriend's eye. A small-town Nebraska teen with aspirations of becoming a vet had bred her chocolate Labrador and was offering the puppies for adoption. The teen nurtured the litter closely, gave them their needed shots, removed their dewclaws. We drove to the rural community to see the puppies for ourselves. A half dozen pudgy balls of joy romped inside a secure enclosure in the teen's front yard. They scampered after each other. Nipped at each other's noses. We couldn't help but smile. One blue-eyed pup dashed in a rumble-tumble up to us, put her tiny paws up on the fence, and wagged her tail like a helicopter. I picked her up and her tongue lapped my face. She was perfect. But my girlfriend and I weren't quite ready to make our decision there and then, so we set her down. The puppy ran back to play with the others.

I was set to deploy to Iraq with my Marine Reserve unit in the next few months, and my girlfriend and I concluded a puppy would make a great companion for her while I was gone, perhaps even better company than me. At least you can train a dog. We headed home, thought it all through, and returned the next day, hoping we could remember the exact puppy we'd connected with the day before. They looked like such little clones of each other.

Sure enough, immediately as we walked toward the garden fence, the same blue-eyed pup sprinted over, put her paws on the fence, and shook her rump. Her voice spoke directly to our hearts: "You're back! You're back! You forgot to take me home yesterday." We scooped her up, paid the teen, and climbed back in our car with our new bundle of joy.

Our puppy's name had to be special. It needed to have meaning, because a dog lives up to the name she is given. When I looked into those big blue puppy eyes standing out against that chocolate brown fur, I was amazed how beautiful she was.

Beautiful.

I pulled out my laptop and searched in other languages for the word "beautiful." Trying my own roots first, I found *spéiriúil* in Irish and *schön* in German. Neither seemed a name that fit, or a name that I could pronounce. So I kept searching. *Linda, hermosa, bonita* — none felt like a proper given name for this tiny being. Then I came across the Italian, *bella.* I looked down at the little pup and asked, "What do you think of this idea?" She chewed on my computer cord, and I exclaimed with a chuckle, "Bella, no!"

A dog named Beautiful.

Bella it was.

Early morning on the day after the consultation, I schedule Bella's amputation. The surgery is set for two days later, so we have a little time left, and that afternoon we go to the park so she can play. It's a sunny day for Nebraska in early May, warm, and the grass is a vivid green from the spring rains. Bella sits for a moment on the grass and I take her picture, knowing it will be her last picture with all four limbs.

Her limbs have always been so in tune with her surroundings — actually, her whole being has. Right from when we brought her home, I found Bella remarkably responsive to training. She learned everything quickly. To teach her to sit, all I did was make the same gesture a few times. I raised my right hand, closed my fist with my thumb and pointer finger outstretched, and pointed toward the ground. She sat her little bottom right down and looked up. "Too easy. I'm sitting. What's next?"

My girlfriend and I took her to puppy classes, and Bella learned how to ignore a savory treat on the ground by hearing the command "leave it," one of the most important things for a dog ever to learn. The command soon worked for anything that needed to be left alone. Cars. Squirrels. A sandwich on the coffee table.

Today she looks intelligent and regal, her head turned to one side, her eyes burrowing forward with intensity and intelligence. She bears no weight on the affected limb due to the pain it causes. Her muscles around the leg are beginning to atrophy. The leg has become nothing more than a painful nuisance. I need to be her voice and take it for her. I walk over, cup her head in my hands, and speak tenderly, trying to explain to her what will happen. "We need to take you back to the vet, Bella. I'm sorry you can't understand all of this, but hopefully you will when the pain goes away." I give her forehead a long and loving kiss.

The surgery will cost fifteen hundred dollars, in addition to the three hundred for X-rays I've already spent. I'm thirty-three years old and in school full-time, finishing a degree in fire protection technology, although I'm still struggling to find my life's true passions. I live as simply as I can in a friend's house that's being renovated. The walls are gutted, I sleep on a cot, and I cook on a hot plate in the basement. To help make ends meet, I sell photographs off my website and work some general labor gigs, supplemented by my military medical retirement. Bella's surgery is no small expense for me, and I'll need to put it on a credit card. *But Bella is worth it,* I remind myself. *I'll pay anything for her, and I'll figure out how to pay for it.*

On the morning of the appointment, I drive Bella over to the clinic, kiss her goodbye, and let her go into the arms of the vet, who she follows with a limp but with a tail that wags and a face that smiles. I don't know too many dogs that love to come to the vet as much as this girl. The whole procedure will take eight hours, and they insist there's no sense in me waiting around, so I head to class and tap my foot, staring at the clock. When class is over, I go home to prepare our room for Bella's return, and then head to a park and pace up and down on a trail. At last my cell phone rings. The vet tries to prepare me for what Bella will look like so I'm not shocked when I see her.

"She's not going to look so good," the vet says. "She'll be shaved, with staples."

I hang up and rush to the clinic. I want to

be there with her as soon as possible. I hope she's not anxious and confused, suddenly waking up with a missing limb with no understanding as to why.

The vet greets me in the waiting room and adds, "It's going to take some time for her to adjust, so you may have to help her walk." She goes into a back room to bring Bella out. I hear a voice from down the hallway: "Wow, she stood right up." A smile grows on my face, as I'm not surprised to hear this. Bella has always been remarkably resilient, and I murmur, "That's my girl." My mind flashes back to years ago when she tore the ligament in her knee at a dog beach and demanded to keep playing while she held one hind leg in the air. "I can still fetch, Daddy! C'mon, throw the toy!"

Today, Bella appears in the hallway, hobbling but already walking under her own power on her three remaining legs. Her tail is wagging, although it's wagging low and slowly between her legs. She looks like she wants to run and play, but she's wobbly from the anesthesia and her eyes don't hold a steady focus. Sure enough, her body is full of staples and it's shaved from the middle of her chest on forward. The front left leg and shoulder blade are gone. It's just like the vet described on the phone, but it doesn't rattle me. I don't see a shaved dog full of staples. I see Bella, alive and well. Kneeling, I open my arms, hug Bella, hold her, and whisper: "You're beautiful, baby girl. Just beautiful."

Bella leans into me, and her tail wags stronger. I've been worried that she might not forgive me for this, that she will be confused, traumatized. As much as I've tried to prepare her for this, I know she can't understand my words — not for something as complex as this. It's difficult. She couldn't make this decision for herself, so I had to make it for her. I've prayed it was the right one. Now, as she leans into me, I can feel that she's trusting the right decision was made for her. Like she's saying, "Thank you for taking that pain away, Daddy." My heart grows warm as I say without words, "You're welcome, baby girl. You're welcome."

I've brought a padded strap with me that's designed to help carry the hindquarters for dogs that have hip problems, and before I can figure out how to adapt it to assist Bella with her front leg, she starts a slow hop toward the door. "C'mon, Daddy, let's go home." I open the clinic door and Bella hops out to the parking lot. She stops at the rear door of the 4Runner and looks up at me as if she knows her limits. "I'm gonna need your help with this one, Daddy." Carefully I reach down and place my arms under her, doing my best to avoid the staples. I lift her into the back of Ruthie, padded with blankets.

"I got you, girl," I tell her. "I got you."

Bella plops down, lays her head down on the blanket, and lets out a heavy sigh. I say, "Yeah, that's it, baby, you just relax." Carefully I drive home, and carefully I put her to bed next to a sign I made earlier that says, WELCOME HOME, BELLA! I wrap her in my old T-shirt from Team Rubicon, the disaster-response organization headed by military veterans that I've volunteered with across the country, and I stroke her forehead, her body, the soft furry place between her nose and eyes. Her eyes close and soon her breathing becomes deep and steady. Her eyelids flutter open and close, like a disturbing dream has come and gone. Then she is resting in a deep doggy sleep. Wiped out by the emotion of the day, I'm not far behind her. I'm hoping that life on three legs will still be one that she can enjoy. I'm hoping that I made the right decision.

At midnight, I hear Bella get up and try to shake. By the time I reach her she is already standing by the front door, a clear signal to be let out. She doesn't ask to go out at night, but her schedule may be off from the surgery, and she still seems a bit disoriented from the anesthetic, so I follow her into the yard. She relieves herself, takes a few steps, then plops down in the grass and immediately falls asleep. It is heartbreaking and adorable all at the same time. I consider carrying her back inside, but she loves the earth, the free feeling of the breeze in her ears. I think: *This moment matters, and it's a moment I will not let slip by.* So I go back inside the house, select a warm blanket for Bella and a sleeping bag for myself, and carry them out. I lift her onto the blanket and cover her with the other half so she won't be chilly, lay out my sleeping bag, crawl inside, and scoot my body next to hers.

She is so close to me. I look at her in the moonlight. I feel the warmth rise from her coat. Carefully, I trace her smooth fur down her spine to her hips. Lightly, I massage down her rear left leg. Gently, I run my fingers around the small growth on her knee. My hand moves up near her jowls, and I wish I could pull on them in jest like she loves me to do. But she desperately needs her rest so she can heal. One of my favorite markings on Bella is a little shapeless brown birthmark on the pink underside of her left jowl. It's her secret little spot that only her closest friends know about, and I long to look at it, but I don't need to see it to know it's there.

I sigh, rest my hand in stillness on Bella so she knows I'm with her, roll onto my back, and look up into the night sky. What a magnificent sight. The vast universe extends out over our heads, and here we are, Bella and I, these tiny creatures below.

In my rawest and most honest moments, in those times when I ask myself *What have I truly built with my life?* I know the only thing I've really got going right now that's *anything* is this dog. And now I'm going to lose her. I'm going to lose it all.

But I'm not going to focus on what I'm losing. Not just yet. I'm going to focus on what we have left, because life isn't over. We have three to six months left, and a quest is forming in my mind. I can't imagine what the quest will look like just yet, but we need to write a beautiful end to Bella's story and change the course of mine.

As I'm lying on my back next to Bella, looking up into the night sky, my thought is this: *Even with all the stars and sky beyond us, right now this three-legged speck is my entire universe. How can this beautiful dog and I make the absolute most of the too few moments we have left?*

2
Let Out to Run Free

We haven't gone anywhere yet. It's been six months now since Bella was first diagnosed with cancer and had her leg removed. She's reached the very end of her projected window of remaining time, but she shows no sign of slowing down, no signs of death. To the contrary, the pain that was evident before the surgery appears to be completely gone. Her head no longer hangs low, and her bright smile is back on her face. Seeing that smile brings a peaceful stillness to my soul. I've made the right choice.

She has grown so adept at moving around on her three remaining legs that I can barely remember her with four. The first time I took her to the lake after she'd healed from her surgery was the first time I ever saw her as anything different from her former self. She bounded into the water after a stick, just as she had thousands of times before. Yet once she got out into the water, she struggled to keep her head above water. With each stroke of her front leg her head bobbed up, but without the other leg to kick in succession, her head dropped to the surface and she took in water through her nose.

Bella had always been such a strong and fearless swimmer. She'd barrel chest-first into lakes, rivers, even the Pacific Ocean, and swim out far enough to worry anyone, then come back to the shore to catch her breath, only to do it again and again. The water was her second home, swimming was her bliss, and to see her struggle with swimming nearly sent me to tears. When she returned from that first swim, she seemed to be a bit confused. "Daddy, why couldn't I keep my head up?"

I knelt down to pet her. "Don't worry, girl, we'll figure this out." I rubbed her wet belly. "At least you're not swimming in circles."

We kept coming back to the lake. The more she swam, the stronger she got. The stronger she got, the more she could keep her head above water.

These past six months have been a bit of a blur. For the most part, Bella and I have made good on my vow to make the most of the moments we have left. We take short hikes and play in parks and swim in lakes and take day trips and weekenders. On many days, in many moments, life is good. But overall, I'm feeling stale, a bit stuck. I've done what I came to Lincoln to do, which was finish the fire protection program, but I had planned on moving on from here. I had big ideas of what I would do once I had my freedom again: travel, adventure. Yet here I still am.

Mostly I'm just killing time, waiting. Bella and I are now living in a rental in Lincoln that I've leased for a year, and in my most honest moments I think: *What are you doing, Rob?* It's a hypothetical question, a lament, usually followed by a negative name-calling session: *Rob. You have failed. Failed at life.*

I know I'm just waiting for Bella to die. I'm her caregiver, and my life's been put on hold until the cancer takes her. But Bella's not to blame for the deep-rooted discontentment I feel. She's the one good thing in my life, and I can't even begin to imagine her not being around. I have tons of ideas running through my mind, all potential solutions to my restlessness, tons of roads I want to travel down to see what's there, but nothing's quite coming together.

One idea that's been percolating for years is to take an actual quest: a long road trip around America. See the remaining states I've not yet visited in this vast country. Maybe I could do it while I still have the world's best copilot with me. I have friends all over, and I'm sure they'd welcome a visit, at least from Bella, and I can tag along. One big question is when. I've toyed for some time now with the idea of waiting to take this trip until after Bella has passed. I've figured it would be too hard for her. Surely it would be best to wait here in Lincoln until she's gone.

But as I picture myself standing on the shore of the Atlantic coastline and looking down and seeing no Bella by my side, I feel a sense of emptiness. That sense prompts me to do this before she's gone. I must do this with her. For her sake, as well as mine. If we wait here, we'll just be waiting to die. I want to experience life, and I want Bella's last months to be lived as fully as possible. We need this journey, and we need it now. But how? We've packed up and left before, but not like this. Not for so long. Not so completely. Not with so many unknowns.

One of my best buds, Avery, invites me to Chicago for the weekend, and he insists I come. It's for the Marine Corps' birthday celebration. I was a staff sergeant in the Marine Reserve at twenty-five, and once a marine, always a marine, although these days, I'm feeling further and further removed from the brotherhood. Avery tells me this short trip will be good for me: I need to reconnect with my brothers from the Corps. As always, Bella will come with me — I wouldn't leave her behind. So I pack a few clothes for the weekend and at the last minute throw in another bag just in case it feels right for us to keep on going.

Bella and I set off for Chicago in the 4Runner. The day is bright, and Bella sticks her head out the window, her nose twitching in the air of the passing countryside. For a while she curls up contented in the passenger seat. Then she rouses herself, shakes, and hops into the back of Ruthie, where I've put the seats down. There, I've fashioned a Hotel de Bella, her own private suite, with a foam pad, blankets, a few favorite doggy toys, and an unspillable water dish. (It took me a couple of trips before I caught on to the unspillable part.)

Bella is a fantastic road tripper, the perfect copilot. She loves riding anywhere in the car with me and stopping for hikes and swims along the way. Over the past couple of years, on other shorter trips, we've already visited the Teton mountains. We've gone down through the slick red rocks of

Moab. We've hiked in the Colorado and New Mexico mountains on trips from Nebraska to California. Ever since she was a puppy, nothing has excited her more than a "r-i-d-e." (Shhhh, you can't say it out loud.) Not only will she fetch the keys on command, oftentimes she'll find the keys and drop them at my feet as her command to me: "Let's go, Daddy!"

Our trip to Chicago feels just like one of those good weekend trips, and soon we arrive at Avery's house in the suburbs. Bella can't go into the city with us, and it won't work for her to stay at Avery's house without me, because they have two elderly cats, a three-year-old, and a newborn, so reluctantly I board Bella in a kennel for the night. Avery and I swing by the airport to pick up our other buddy, Pete, then all three of us head downtown. It feels good to be with them. Distance separates us as brothers, but we try to get together at least once a year.

The birthday celebration goes off without a hitch. Still, I'm distracted with visions of Bella in a dark kennel. She's done great being kenneled in the past, but our bond has grown so much stronger over these recent months. I've rarely left her side since the surgery. Maybe she's wondering where I went or if I'm ever coming back. First thing next morning I pick up Bella from the kennel. Rather than rush to greet me, she rushes to the rocks in the parking lot outside and pees more than I've ever seen her pee. My jaw clenches and I wonder if they let her outside at all. Though I've had good experiences before with kennels, I vow that this will be the last time I ever kennel her and say, "I'm sorry, baby girl. I won't do this again. Ever!" She barks at me, licks my face, and pulls toward Ruthie as if to answer: "I'm upset that you left me. But I'm already over it. I love you! Let's go for a ride!" Never one to hold a grudge, Bella hops right into Ruthie, and all is forgiven. If only I could learn to forgive so quickly and move on.

We are invited to another buddy's house in Chicago, and I realize I've got a great network here. My friends want me to stay, to stop wandering, to stop searching. They offer to plug me in their networks. But Chicago is not the answer. Bella and I could settle here, get some job just to work a job, but we're basically in the same place in life we were in in Nebraska — still searching for the elixir, still leaving Bella at home while I'd go to work as dreams fade away. I need more than that; Bella deserves better than that. I thank my buddy for allowing me to spend time at his place, pack our things back into Ruthie, load Bella up in the back, and take a seat behind the wheel.

That's when I see the extra bag.

That extra bag I'd packed, "just in case." Just in case it feels right to keep going, keep adventuring. Perhaps the elixir is as simple as that. Bella and I, living our grand adventure. It's what makes both of us the happiest, and that's what this should be about. As Bella and I head out of the city in Ruthie, I feel there's still a dutiful tug for us to return to Nebraska, but the road today is somehow different than we've seen it before. It shimmers before us, beckoning, and I notice anew how solid and honest the steering wheel feels in the grip of my hands. The sky is clear and cold, and the road stretches ahead to the horizon, and I think, *This is it.*

Our moment of decision.

Here and now we must take the lead in our lives. In this moment, we must stop holding our breath on what our heart says and start going instead with what our gut feels. These are *our* lives. Our problems to solve. Our adventures to take. It's time for me to give my girl what she deserves. Time to set forth on a journey that's ours to own and behold. It's time to stop dreaming and to start doing.

Bella sits in the back seat with her head out the window. Her ears flap and her jowls gently flutter in the wind. I can see her in the rearview mirror, and my eyes focus on her for five seconds, ten. Twenty. It feels like an eternity while driving. My eyes focus back to the road ahead, and I suddenly feel at peace with this decision. I know I need to take this grand trip with Bella, and I need to take it right now. We don't have much money, and we don't know where we're going to stay, but I know this is what we need, this adventure. I also know we need to pay attention to the lessons along the way. We need to do lots of thinking, have lots of conversations, take lots of photos, write it all down, grow our perspective, and maybe come to some conclusions in our journey. Most important, we need to simply be with each other until the end.

I figure we'll zigzag all over the East Coast and down to Florida, then maybe home to Nebraska for a while, then head west through the heartland toward California, and then up the coast to the Pacific Northwest and finish up in Oregon — suddenly I can visualize the route in my head. Bella loves seeing new things as much as I do. When we reach our destination, we'll go on hikes in the lush green forests that I've seen only in photographs. Bella can swim in the Pacific Ocean as I take photos of the iconic coastline — and then we'll have done it. We'll have completed our final journey, having explored the country side by side. We'll know we've made the absolute best use of our time left together. Then we can say goodbye.

Our decision is made. Every detour, every wrong turn, every pothole, and every roadblock has led us to this moment, because in this moment we are alive. We have everything we need to survive, and as long as we are alive, we will aim to live, together, until the end. We will take a breath, look around, and live here, now, today. Goodbye, society. Hello, *life.*

Shit. It's only ten miles later, and I don't know where we're going. I don't know what we're doing. Maybe we should turn around and go back to Nebraska.

But isn't this the whole point? Living with this feeling, grappling with this tension. We haven't figured everything out — not on this trip, and not in life either. The journey itself becomes the destination, and what we need to do from here on out is stare at the GPS and pick a route on the map. The beauty of having no plan is discovering destiny along the way. How many important serendipitous moments are missed because those moments are not in the plan?

As we keep driving, I decide to go north, because it's easier to camp with Bella in cooler weather. If she gets too hot, it's harder to cool her off. So we'll try to stay away from the South at first. The tops of the hills are already turning white, and whenever we stop and pull over at a trailhead or a park, we're the only people around.

I think: *Maybe that's the way it should be for a while.*

We visit friends in Grand Rapids and head into Detroit and see forgotten neighborhoods filled with blight. I've read about the dilapidated state of this once great city, but seeing it in person reveals the palpable reality of its degradation. Several houses have fire damage, some with their entire roofs burned away. Surrounding houses have their windows cracked and shattered. Yards are completely overtaken by unpruned trees and waist-high weeds. Entire neighborhoods remind me of disaster areas I've responded to. There's a heavy energy in this part of the city, and Bella's mouth is closed. She's breathless, not even panting. I see how this stop in Detroit has significance in our journey.

We meet up with a fellow volunteer from Team Rubicon, and he invites us to stay overnight in his loft. The next morning, Bella and I tour a few of his jobsites, and he explains how he's working with a local organization to help restore the city, house by house, block by block. He wants to make the city beautiful again. It's such a severe problem, this city once so bustling and full of industry. Jobs went away, people moved out, and destruction set in. Now there are large tracts that so desperately need restoration. It dawns on me that Detroit is actually a metaphor for the start of our journey. Friends were lost, opportunities fell away, and the taste of blight runs heavy in our mouths. Restoration is needed. New purpose. New vision. We can choose life, or we can choose death.

Bella and I say goodbye to our friend, head into downtown Detroit, stop, and get out in front of a graffitied mural of a giant shark, its mouth gaping and toothy. I take a picture, and Bella is sitting up, straight and alert, the muscles of her lone front leg strong and defined. Behind her stand several abandoned buildings brought back to life with vivid colors and brilliant designs. I've seen graffiti before, but nothing like this. We've been told that artists travel from around the world for the opportunity to transform these brick canvases. The city has actually sanctioned it. City officials didn't give up on these buildings or tear them down. They allowed the world to create something beautiful from them. It makes me think how I haven't given up on Bella even when she was damaged. I didn't end her life when many said I should. Now here she is, standing boldly before me, once broken, now stronger than ever — and she's not giving up on me either.

Many people say they love dogs more than people, and I always respond, "Give people a chance!" I do have a bit of a preference, though, and that is that I do like dog people more than non–dog people. They're my favorite people; we get each other. I'd like to say that I'm very open-minded, but I will make a judgment when someone says they don't like dogs. Also, within the realm of dog people there are still many different types. Some treat their dogs as a tool to help with work, and that is that. Some treat dogs as humans without teaching them any boundaries or discipline. I'd like to think Bella and I fall in the middle, but I definitely lean more toward Bella as a human. Anytime someone refers to her as "it" or "the dog," I always interject . . . "Her name is Bella."

"How could you spend so much money on a dog?"

"It's just a damn dog."

"You should just put her down."

These are things that dog people never say.

I set a timer on my iPhone and take another photo, this time of the two of us in front of an old loading bay. When I retrieve my phone, I see something I hadn't yet seen in the city. I see myself. I'm wearing the same black fleece jacket and cargo pants from thrift stores and my hiking shoes and Carhartt stocking cap. Yet my head is held higher than it's been in a while, my shoulders are back, my arm is around Bella, and my camera is around my neck. Bella's face looks confident, and my face looks confident too — and it hasn't looked that way for a long time.

Bella and I drive out of the city on Eight Mile Road, the main segregating line between rich and poor, and that night we sleep in the parking lot of a twenty-four-hour fitness center. Bella sleeps like a rock, yet although it feels safe, I find I'm hypervigilant. It's the first time we've both slept in the car in a city. We're tucked away securely in the 4Runner, but the sound of any distant shutting door causes me to open my eyes, peering through the darkness to see if it's someone breaking into cars, or a security guard telling me to move on. I watch Bella on and off throughout the night. So peaceful. So secure. I envy her. Oh, to be as content as a dog. We sleep just enough until daylight, and then we start driving again.

We drive on to Bowling Green, Ohio, and stop to visit family friends. December has arrived and the weather is frosty. Christmas is already in the air, and as Bella and I walk down the street we admire the remarkable attention to detail seen in the construction of the buildings. We hear the bells of the Salvation Army ringers and I swear I smell freshly baked cookies in the air. Bella hops down the street on her three legs, and people stop on the sidewalk to admire her bright eyes and wagging tail. "Inspiring pooch!" a man says, and nods toward her coat. "She's a real example of perseverance." Bella smiles, and the man smiles back at her, and I see how Bella's happiness is so contagious. Maybe that's her purpose. She's more than simply my copilot. She's spreading happiness and inspiration wherever we go. Maybe that will become my new purpose — introducing Bella to more people, showing how much this dog with three legs and terminal cancer still loves life.

Bella and I learn about Put-in-Bay, a tiny island community off the shore of Lake Erie, about eight square miles, with a winter resident population of no more than a hundred and fifty folks. Right away it intrigues us. During the summer, Put-in-Bay becomes a party town, the Key West of the North, and the population swells to maybe fifteen thousand barhopping vacationers. But now, in winter, it's desolate. Virtually abandoned. It's a town where life ebbs and flows. Bella and I simply must go. We drive Ruthie to the ferry terminal, and the ticket guy asks, "When you coming back?"

We shrug, and he explains there's a time frame with ferry crossings. If we cross now, then we won't be able to return until the next day. We see on a map there's a state park on the island, so I know Bella and I can always car camp there. We drive Ruthie onto the ferry and park, and then the ferry fires up and starts moving. It's the first ferry ride ever for both Bella and me — and Bella has a quizzical look on her face. She's still in the car. The car's stopped, but she's still moving. "Daddy, why is the parking lot moving?"

Off the ferry, the landscape on the island is eerily dotted with a miniature-golf course, a go-kart track, and a long row of bars. Everything is shut down for the season, and it looks like a ghost town. Bella's tail wags. "Just let me out, I can have fun anywhere!" I'm sure she would love tearing through the mini-golf course, but we'd better keep driving to see what we can find on this little hamlet of land.

We loop counterclockwise around the island's perimeter and find ourselves at South Bass Island State Park. It turns out that the community is named Put-in-Bay, and the island itself is named South Bass. The back-door windows of the 4Runner are down, and Bella races to and fro with anticipation. She senses she's about to be let out to run free. We wind through the camping spots and find the entire park to be empty. I park Ruthie near the main grassy area of the park, open the rear door, and say, "Looks like this park was made for you and me, baby girl!"

Bella jumps out, runs a quick lap, then rolls in the grass. "Yup, this is perfect, Daddy!"

Perfect. I think about that word for a

minute. Of course, this experience is perfect. Bella doesn't need go-karts or miniature golf. This little field of grass will do just fine. Bella brings the life.

She finishes her playful romp, then hops alongside me as I walk toward the shoreline. When we reach the edge of the grassy field, we look over and down, and it turns out we are on the top of a sheer cliff. The land towers above the giant lake below us, Lake Erie. From this perspective, it might as well be the ocean. Along the edge, I keep an eye on Bella for safety's sake, but I'm amazed at her fearlessness. We're about two stories above the shoreline, and Bella surveys the scene mere inches from the edge. Her ears float in the air as the wind passes beneath them. The sight is magnificent, and in this moment I can't help but think how in love I am with my beautiful traveling companion.

"You wanna get down there and check it out?" I ask.

Bella's mouth closes, her ears perk, and her head tilts to the side as she focuses on my next move. She's not sure exactly what I said, but she knows the words "you wanna" are always followed by something fun, and fun is what this girl lives for. "Yeah, Daddy, yeah!" she seems to say as she wags her tail and begins to hop up and down.

We find a short switchback that leads us down to the water, and Bella navigates the steep grade like a champ. The shore, however, proves to be trickier terrain. It's comprised of relatively large, flat, smooth rocks with rounded edges. These make it difficult to walk on, particularly for a dog with three legs, but as I bend down to inspect the rocks more closely, I see they'll make perfect skipping stones. I sort through a few to find the best one, and in my mind, I'm suddenly transported from this Great Lake to a small Nebraska lake from my childhood. It's Lake Maloney, just outside of North Platte, and I'm about seven years old. My older brother Mike and I are walking along the shore of the beach, scouring for the perfect rocks to skip. Mike and I shared a bedroom for the first ten years of my life, and we often played together, building tent forts in our room and snow forts in the backyard. He made tracks in the dirt for my Matchbox cars. He was the one who taught me how to play catch. He taught me how to skip stones.

"You need them to be flat and smooth," Mike calls to me.

I keep looking on the ground, but most of the rocks I see are round and jagged. Useless.

Mike soon finds a great one. "Here, use your wrist like this," he says, and whips his elbow through the air toward the water and lets go of the rock.

I count. One . . . two . . . three . . . four . . . five. Five skips. Five separate rings ripple into each other. Wow, I want to do that!

"Just keep looking," Mike says. "If you look hard enough, you're bound to find what you're looking for."

I search and search and finally find a rock of my own. The rock is not perfect, but it'll work, I think. I give it a toss toward the lake, trying to replicate the action Mike just showed me.

I count. One . . . *kerplunk*. One giant ripple.

"Gotta use more wrist," Mike says. "Here, watch again."

We skip rocks for hours. Him teaching. Me learning. His rocks skipping. My rocks plopping. I try and try again, constantly searching for the perfect rock, constantly trying to get my throw just right. Then, as the sun is setting, I find one last perfect rock. I whip my arm out from my side and let the rock go with a sharp flick of my wrist.

I count. One . . . two . . . three . . . four . . . five . . . six!

Six!

"You did it, lil' man!" Mike says. He comes close to me and ruffles the hair on my head.

The vision vanishes as quickly as it appeared. Bella and I are on the shores of Lake Erie again, and I'm smiling a soulful smile. A handful of flat rocks come into focus. I stoop down, pick one up, and look up to the big blue sky.

"How's this one?" I ask, still looking up into the air.

Bella is by my feet now. She stares up into my face, her eyes inquisitive. "Who you talking to, Daddy?"

I pause, wipe away the wetness at my eyes with the back of my hand, look at Bella again, and smile. I whip my elbow and release the rock with a sharp flick of my wrist, just like Mike showed me. The rock hits Lake Erie and takes two giant skips, followed by a trickling of too many skips to count. I put my hands in my pockets, satisfied, and watch the ripples form into each other, then smooth away as if the surface were never disturbed. I kneel and hug Bella. She puts her head into my legs and I rub her ears and shoulders.

"C'mon," I say simply, and straighten up. Bella hops forward, and she and I find our way back up to the grassy area overlooking the cliffs, where I get us a blanket from the 4Runner, along with some snacks — peanut butter and crackers for me, a handful of doggy treats for her. Together, we sit and watch the sun go down. I've taken so many photos of just Bella, I decide that I want one of the two of us. I want to save these perfect moments so I can remember them forever. I'm about to lose my best friend, and I know that feeling of loss and how awful it is. At least I have some warning with her. If I didn't know her life had a time limit, then perhaps I wouldn't be sitting here appreciating the magic of this sunset.

Bella hops over and licks a smidge of peanut butter caught in my beard. I chuckle as I hold on to her. The same sun has risen and set every day of my life, yet I feel as if I'm seeing it for the first time. As the shadows of nearby trees grow longer, worries about Bella, myself, and the world are swept away by the gentle breeze. The sun sinks closer to the horizon, sending a reflection through the water below us that appears to be a golden road. A road leading straight to that place beyond. My heart and mind blend together and life becomes clear. We are alive. We have experienced the gift of life another magical day, and for the first time in a while, I am grateful for that simple fact. This is truly all we need. We don't need more money or things. We all need more of this.

Darkness takes over as the sun disappears behind the horizon, and as I hold Bella close, my heart smiles in the form of a deep, happy sigh, knowing that the sun will rise again tomorrow and, for now, we'll both still be here, still alive. Still living fully.

3
Only One Problem — Me

Before we got Bella, my girlfriend and I read a few handbooks on raising puppies and learned about the benefits of crate training, but like many first-time dog owners, we weren't sure if crate training was right for our puppy or not. At first glance, it felt like punishment to keep a dog locked up in such a small space for hours at a time. Yet the more we read and studied and the more we talked to friends who had experience with dogs, we came to see that dogs are natural den dwellers. Providing them their own safe space is actually a thoughtful practice. So we decided it was the best route for Bella.

The first night after we brought her home, we got her safely stowed away in her crate and reassured her with low, gentle voices. Yet as soon as we left, Bella let out the most sorrowful whine. It took all I had to not go in and rescue the poor soul and cradle her pudgy brown body. My girlfriend reminded me that enduring a few nights of whining would all go toward the greater good.

"It's okay, Rob," she said. "Bella needs to learn to cope on her own. We won't always be able to be home with her."

I buried my head in the pillow and tried to drown out my new baby's calls of distress.

When I woke the next morning, I rushed to let her out of her solitary confinement. Bella was nestled quietly on her little doggy bed, sound asleep, contented and relaxed. She woke up, yawned a lazy yawn, and eyed me lovingly. "What's for breakfast, Daddy? Let's play!"

As the weeks went by, she quickly learned the command "kennel up" and trotted into her crate without dispute. Not once did she pee or poo in her little safe space, and she never seemed as if she were being punished when we closed its door.

Weeks turned into months, and we concluded that the crate training had paid off. Whenever we left the house without her, we knew she was safe. She couldn't get into any trouble around the house. She couldn't chew the furniture or eat some random chemical we didn't even know we had in a cupboard somewhere. This knowledge gave us the confidence to leave her at home and know she'd stay safe.

One afternoon, I realized just how right this decision was as I found myself shutting the door to her empty kennel because I wanted her to stay in the living room with us and snuggle on the floor as we watched a movie.

"I'm glad you like your safe space, baby girl," I said. "But don't forget you'll always be safe out here with us too!"

Growing up for me felt a bit like crate training, although I'm not positive everything I went through was for my greater good. There was more than one time I felt lonely, more than one morning I woke up and didn't feel safe and contented. All these experiences became part of who I am today. They translated into the fodder of my heart that I need to process as an adult. And I know I'm not the only one. If I've learned one thing from my short time on this earth, it's that everybody has a story. It's what has allowed me to have meaningful, open-minded conversations with people from all walks of life throughout the years. Everybody journeys through deep and narrow valleys in search of a land of kindness, a land of safety and belonging, a land called home.

Lately, my time has been packed with meeting so many people from different origins and various paths. I've seen that while people might disagree about beliefs and ideals, when we take the time to learn each other's stories, we see that no matter our differences, we also have a lot of similarities at our cores, and these similarities allow us to relate to each other. When we look for these similarities, we see others for who they are — what they've been through, who they've loved and who they've lost — not what they have accomplished, not what they believe, or how much money they bring home. The relatable things that translate into every language all involve this fodder of the heart. These relatable things are love, loss, heartache, and passion. These reside within all of us, regardless of race, origin, or creed. These similarities allow us to put aside our differences and prejudices and grasp this vital truth: that you love and hurt just like I do.

So to understand someone's story, we must listen to the heart of their origin. My own story started long before Bella came into my life. It started with a broken home, a home without any dog, even though I longed for one. I know now that when I was a child, a teen, and even a young man, when I experienced loss and heartache, I was actually longing for this land of acceptance, this land of love and belonging.

I was born into a blended family, the Kuglers and the Dohenys. A family where I knew I was loved, but also a family where I never felt like I fully belonged. Both my parents had previously been married and divorced, and each brought three kids into the new marriage. Then, in 1982, Mom and Dad had me together, the baby of the whole bunch, so I was never a full biological sibling with anyone, although I grew up with six older brothers and sisters: from my mom's side, John, twelve; Amy, seven; and Mike, five; and from my dad's side, Charity, ten; Joy, eight; and Jason, four.

We were like the Brady Bunch plus one, although Dad wasn't a successful architect and Mom was never a picture-perfect housewife. We certainly never had a cheery housekeeper named Alice. For a while we lived in Stapleton, Nebraska, population 305, in two trailers connected with a make-shift plywood hallway. My dad, Myron, drove a school bus, and my mom, Kathy, stayed home with all seven of us kids. It was a recipe for failure. It wasn't long before one trailer ended up being the Dohenys' and one trailer ended up being the Kuglers'. Myron and Kathy divorced before I turned three, and the two biological families went their separate ways. Mom's kids with her. Dad's kids with him.

Only one problem.

Me.

Where did I belong? There was a custody battle, and Mom won, with Dad keeping me summers and some Christmases and Thanksgivings. I don't remember the battle, but I do have a memory that plays like a movie: I am looking with teary eyes through the back window of my mother's car, reaching for the blurry figures that are my father, brother, and sisters as we drive away.

I can only imagine the pressures of being a single mom and raising four kids. After the divorce, Mom scrambled to find a way to support us. Though she had a degree in dental assisting, a hearing impairment she'd had since birth prevented her from ever being able to work in the field. She got by in life by reading lips, but lips can't be read from behind the surgical masks dentists wear. Once she became a mother, she found herself needing to work her first full time job, at forty-one years old. Rather than give up or ask for help, she became a Certified Nursing Assistant. Unfortunately, she slipped a disk in her back while lifting a patient out of a tub, ending her nursing job, so she worked any job she could find to make sure we didn't go hungry.

Dad remarried and moved to Colorado to attend seminary school, and his presence in my life became distant, allocated only to summers and every other Christmas or Thanksgiving. He still had three kids to raise and could send very little in child support, although he sent whatever he could. Things weren't easy at Mom's house. I was a Kugler living with Dohenys, though Mom kept the Kugler name so I wouldn't feel alone. The first house we lived in was scheduled for demolition, but Mom persuaded the owner to let her rent it. The floor was so slanted that when my brothers and I played with our Matchbox cars they would turn and run into the wall. Each morning our first task was to check the mousetraps in the silverware drawers. I remember being excited each time we caught one. When the cold Nebraska winters hit, the house was heated by a woodburning stove that my older brother John cut the firewood for. My sister slept in a built-on room that had mold and moss growing on the cement walls. Not able to afford snow boots, we put plastic bread sacks over our socks before we slipped them into our shoes. Mom crocheted our hats out of yarn. I didn't know any different, until I started going to school. There, I saw that the other kids not only had snow boots and jackets, but their entire lives seemed to be different, too. They went on vacations and jaunts to Disneyland, while the few times we traveled, we went to funerals. Only twice did we go on food stamps, which Mom hated, because she never wanted to be seen needing help. I admire her for that now, but we relished food stamps as kids. Food stamps meant full cupboards, more than the usual cereal and cheap lunch meat.

Mom was a good mother in so many ways. She always loved us. She never drank or smoked or did drugs. She had an undying work ethic, and her priorities were bills and food. And she had a strong moral backbone and a fierce Irish temper that us kids inherited as well. She also taught us about living simply and having compassion not only for people, but for animals, often taking in cats that needed a home. I'll always remember Muffin the sassy tortoiseshell and her son, a black behemoth named Pudge. They became members of the family and gave me my first glimpse of what it meant to care for someone other than myself. Dad, who I visited in the summers, worked as a building manager for a school in Colorado to pay his way through college to become a

Nazarene minister. He and his new wife, Donna, a nurse, were strict about their religious beliefs. Our free time was spent memorizing Bible verses, and we couldn't say "geez" because it was too close to "Jesus." Nonetheless, their intentions were from the heart and their teachings instilled morals and a sense of service to others. Donna had a son named Tony, twelve years older than me, who adopted me as his little brother.

The love from both families was evident. I never questioned that. Yet with both families I always felt like an outsider, the solitary full-biological son of Myron and Kathy who needed to divide his life between Nebraska and Colorado, with no one place to call home. When I was in fifth grade, Mom scraped together enough cash to put a down payment on a small house in Broken Bow, just one of many small farm towns along the railway tracks west of Lincoln. Maybe this would be it. Finally, a place for me to belong. I was equally excited because having a house meant we could have a dog — right? Surely, a dog would help.

My excitement grew. Not long after we moved in, I was walking to school one morning and spotted the perfect Lab–Border collie mix in a neighbor's fully fenced backyard. She'd just had a litter of puppies, so I carefully approached the fence with an open palm. The mother dog approached me from the other side and sniffed. I guess I checked out okay, because her tail wagged and she lay down on the grass, and her puppies scampered up to the fence and stuck their little noses through the slats. I reached through and petted the puppies and they nibbled on my fingers with their sharp little teeth.

The next morning, I stopped and did the same thing, and the morning after that, and the morning after that. Eventually, I narrowed down all the puppies in the litter to my three favorites: the all-black one, the black-and-tan one that looked more shepherd, and the black one with the white tail. The man who lived there saw me one day and said hello, and I asked him if the puppies were for sale. He shrugged good-naturedly and said the dogs were just mutts; he was going to give them away. Did I want one?

Did I?!

"You can have any one you want," he said. "Just as long as it's all right with your mom."

This was my chance at a puppy. My puppy.

After school that day I raced home, my speech to Mom prepared. A dog was a needful part of boyhood. A dog would teach me responsibility. A dog would help grow me up. I needed to wait until Mom was home from work. I was so hoping she'd say yes. But when she came home, she gave me a hug and looked around the kitchen with weary eyes, probably wondering what to do for supper. Before I could even tell her the fuller story, the answer was already no.

No, Bobby. Puppies are a lot of work.

No, Bobby. We don't have a fence, and a dog needs a fenced yard.

No, Bobby. You can't build a fence yourself.

No.

We ate supper that night, and I went to bed and the next morning walked by the yard again to see the puppies. Day after day followed, and I just kept walking by the yard. Still staying hello. Still looking. Still longing.

One by one, the puppies disappeared.

I think you have to understand what you *didn't* have to fully appreciate what you *do* have later on when it's good. No, I never had a dog while growing up. But yes, I had such an incredible dog in Bella later on. And

I became the right person for her, too. Somehow my growing-up experiences transformed me into a person who understands commitment, someone who must feel a greater sense of commitment in those closest to him, a steward of those he cares for.

Perfect qualities for a dog person.

A dog needs a good steward. And a good steward needs a dog. Because people who love dogs know you must *be there* for a dog. A dog can't just come and go like a cat. You must commit all the way to a dog. And when you are there, and you let this attitude of being truly committed and truly present in each moment permeate the rest of your life, well, you're onto something really great then.

Fast-forward to when my girlfriend and I first got Bella. We knew right away that we needed our dog to be able to go everywhere with us. Some people are animal people to the point that they'd rather spend their time with animals than people. Me, I prefer to do both at the same time. I'm a people person and have a very large friend circle that's an integral part of my happiness in life. To make sure Bella could do well with people, it was important that we started her socialization as soon as her vaccinations had their chance to work and it was safe to bring her out and about in the world.

When Bella was barely three months old, her little belly was still round and her eyes were just starting to lose the blueness and switch over to a copper color. A friend was hosting a barbecue at his place, and I asked if we could bring Bella. "Of course, man! I'd love to meet her!" he said. So we loaded up the car, and Bella sat at my feet on the floor of our Honda Accord while my girlfriend drove. I scooped Bella up, holding her closely, and allowed her the chance to sniff out the passenger-side window. The rushing air tickled her nose, and she let out the most adorable little sneeze as her tiny ears fluttered in the wind.

We pulled up to our friend's house. Most of the people were congregated in the garage. I set Bella down in the driveway and she stuck beside me for a few steps, then ran toward the people, introducing herself and sniffing the curious aromas they'd collected on their shoes. She sniffed her way toward the back corner, the back corner every garage seems to have, with tools on the shelves and a bit of oil spilled on the pavement.

"Nuh-uh! Bella, come," I said.

She stopped and came back to me, her tiny tail wagging.

My buddy voiced wonder: "Whoa, she already listens?"

"Dude, she's so awesome," I said. "Here, watch this." I gave Bella the command to sit, and she sat down and kept her eyes trained on me.

"Good girl!" I bent down, scooped her up, cradled her, and rubbed on her belly as she bit at my nose with her tiny puppy teeth.

"Man, she's gonna be a great dog!" my bud said as he gave me a pat on the shoulder.

"Already is," I said with a smile.

In those early years for me, nothing much was great. I was often bullied for being small and poor. My grades were usually lousy, and I found it hard to concentrate, probably due to undiagnosed ADHD. We didn't have the financial means or health coverage to see therapists, so there was no diagnosis until adulthood. An inhaler was a constant companion, making me an easy target. Over the years I encountered different bullies, different incidents, but when I was sixteen I decided to do something about it. I hit the gym and started packing on muscle. My motivation was part self-preservation, and part revenge. I imagined smashing in the heads of anyone who'd ever bullied me, so I pushed the weights faster and harder. My physique filled out, and my homeroom teacher remarked, "Kugler, what are you eating? You used to have noodle arms, now you have guns of steel." I went from 150 to 175 pounds and ended up one of the strongest kids in my class. I gained the confidence to join the football team, but muscle doesn't help you learn the game. I spent my summers with my dad in Colorado, missing out on the football camps that teach the basics. So, without understanding the game, I ran around wildly, simply trying to hit the guy with the ball. I was on the team, but I spent most of my time on the sidelines. I wanted to try out for track, but Mom would only approve one sport per year. We needed to spend the rest of the time working, earning our own way. That was the emphasis. Work. Bills. Food.

Then a dream took hold: joining the military and serving our country. My brother Mike was my inspiration there. I had seen how it had molded him as a man, how he became a part of something bigger than himself.

Five years older than me, Mike was just a teen himself when he stole some toys from the tractor supply store he worked at. He stole the toys so he could give them to *me.*

Stealing is nothing either of us were proud of; I'd done it too. We worked plenty too. Don't get me wrong. We worked our asses off, even as kids. Mom showed us that. She ingrained in us a work ethic and showed us how to stretch a dollar, but sometimes we wanted what the other kids had — so sometimes we stole. Mike got caught, and he'd been in trouble before, so the court stepped in, and Mike was sent to a youth rehabilitation treatment center — a no-nonsense, boot camp–style facility that proved a wake-up call for him.

Mike worked hard at the center and straightened up any wayward ways. When he came home, Mike was mentored by an older family friend and rancher who'd been a paratrooper during Vietnam. After a summer on the farm my brother decided to join the Marines. It was his Christmas present to my mom, a benchmark of his new ways. A few months later, we sat in the bleachers in San Diego and watched him graduate boot camp. He had proven himself — and also set the standard. I knew then I would follow in his footsteps.

With my own father living in another state, and both grandfathers passing by the time I was six, Mike became my mentor, the man who would step up and fight for me, the man I decided I would mold myself after. We had our disagreements, but they never lasted long. He loved me. He didn't always like me, but he completely loved me.

He became the one constant I could always count on.

Man, those early days with Bella. So many memories of life lived flat-out. Flat-out beautiful. Flat-out fun.

Within a few months of being born, Bella transformed from a pudgy, tumbling ball of chocolate into a sure-footed and athletic dog. Her limbs grew longer. Her body began to take shape. She truly came into form. Yet one day when Bella and I were playing out on our back porch, I realized she hadn't quite understood the purpose of her tail yet.

Bella contorted her body side to side quickly, staring this way and that at her wagging hindquarters. She bent her body like a horseshoe, still wagging her tail, inching as close as she could get to this enigma, when suddenly she caught it and held on.

"Well, you got it!" I said, laughing. "Now what are ya going to do with it?" I'd seen a lot of dogs chase their tail, but I couldn't recall many actually catching theirs.

Bella stood still for a split second, her tail

in her mouth, her mind undoubtedly a bundle of questions, then decided to run off with her new prize. She jumped, hopped, ran, tumbled — all with her body still bent in half — then collided into the lattice on the porch. Her tail slipped out of her mouth. She gave me a quick look of utter disappointment, then lunged for her tail, grabbed it again, and again took off running, jumping, tumbling, refusing to give up the chase. This time she bounced off her kiddie pool, hit the side of the house, then crashed back into the same lattice. The ability to stay determined when others doubt you is a great trait. I was sure Bella was going to do just fine in this world.

"We've all been there, baby girl!" I called out with a grin. "I've had to pull my head out of my ass more than once. But we learn our lessons!"

Bella's antics were kind of like my growing-up years. I bounced off plenty of walls and hit plenty of lattices, and although some of it might have looked like slapstick from the outside, it wasn't all fun underneath for me. With the constant traveling back and forth between my mom's house and my dad's house, the bullying, the always scraping by financially as a family, the feeling like I never fully belonged, a lot of seasons went by where I felt like I was running sideways with my tail between my teeth. Along the way, a cog was set in motion deep in my heart that developed into a searching, a longing, a propensity to wander. I often didn't want to be where I was in life, so I concluded that if I could just go someplace else, become somebody different — finally catch my own tail — then maybe life would make sense.

4
Everyone's a Friend

It's the end of 2015 and the road trip is still young. The peace of Put-in-Bay is not long behind us, but already I'm wishing for more of it in my life. Bella and I are on the road again, heading east toward Cleveland. Ruthie feels good on the highway. Bella's eyes shine brightly in my rearview mirror, and I'm musing about how friends insist I've developed three personalities over the years. There's the fun, likable me, a leader who pursues what he wants in life while looking out for others. Then there's the pissed-off guy. Even uncontrollable. This guy wants to punch you in the face for looking at him wrong. Then there's the guy who's super depressed. Mopey and lethargic. Nobody wants to be around this guy. Well, maybe a really forgiving dog, like Bella, doesn't mind too much. She seems to accept me, whoever I happen to be at the time.

Through the windshield I glance up and see the winter sun peeking through the clouds. The pissed-off me and the depressed me have shown up too much in the last few years, and my eyes snap back to the road and wonder when the nice guy will show up to stay. I'm mostly happy-go-lucky, yet my explosive side lies tucked underneath that personality. And my mopey side? Well, everybody has a bit of mope. Me, some days too much.

I put these thoughts aside as Cleveland rises toward us, and we drive downtown, park, climb out, stretch, and head out for a walk through the tall buildings. Bella is playful, her tongue panting, her tail wagging, happy to explore even the concrete wilderness. It's not all concrete, as we find a nice green lawn among the cityscape. Bella brings the life again and rolls on her back in the green grass, and I snap a photo to capture her joyful play in nature juxtaposed with the concrete jungle around her. She can make me a better me, and she can make anything a better photo. Downtown Cleveland is busy, noisy, with people everywhere and thick traffic. As we return to Ruthie and get set to leave the city, we are caught in a construction zone. A cool-looking building invites a photo, so I pull over in a closed lane, hop out, and click the picture, but as I'm leaving I hear: "Hey, come here!"

The voice is gruff. A police officer glares in my direction, motioning me to approach. Bella stays in the car, watching me. Before I can respond, the police officer says, "Show me some ID. What the hell do you think you're doing?"

"Just taking a picture," I say. My heart beats faster. Bella pops her head out the window to investigate.

"Tell me why I shouldn't arrest you right now."

I'm not sure exactly how to respond. I glance at Bella, who's observing with a wagging tail. Got to play this cool. I flip open my wallet to reveal my driver's license in a plastic window next to my military ID. I never hand an officer my military ID, knowing he or she may take this as a ploy, but I've learned there's a subtle art of simply letting a cop know it's there. It sheds a bit of light on my backstory and lets him know that, at one point in time, I did my part, like he's doing his.

"Don't be an idiot," the officer snaps before I can say anything more, and pushes my ID back toward me. "This is a busy street. Don't be stupid."

It all clicks in a flash. His demeanor. His

tone. He's king of this territory, and I'm breaking his rules. I'm in the wrong, I need to leave, and arguing about his tone will do me no good. My quickest escape is to say "Yes, sir. Sorry, sir" and go about my merry business — and it doesn't mean I'm giving in. I'm just playing it smart. You have to know what battles to fight, and I have no need to save face in his one-street kingdom; I have an entire country yet to explore. So I quickly comply and jump back into Ruthie.

Bella sighs with satisfaction. Her gaze wanders, taking in the random sights of the city. She's okay. But I'm not. As we drive down the street, searching for another place to park, I still feel a flush of anger as I replay the exchange in my mind. The officer's just doing his job, trying to keep the city flowing, and this is his block. I get that. I respect law enforcement officers and their duty to protect and serve. In fact, I have many friends who wear the badge. But shit, you don't have to be an asshole about it. You don't have to treat people like they're inferior. I can still see his face, hear the annoyance in his voice, taste the condescension he spat in my direction. Technically I was wrong, but threatening my arrest for such an insignificant infraction seems to be flexing the power more than keeping the peace.

I let it get to me. My mind flashes to the childhood bullying, and suppressed memories surface, playing the situation over and over in my mind until I boil over. I punch the steering wheel and let out a raging shout. “GAAAAAH!”

Bella isn’t scared by this. In fact, she knows exactly what to do. She’s done it a thousand times before. She gets up from her nestled blankets, hops her one front leg onto the center console, and leans into me. No face licking, no anxious wagging tail. Just a slow, still lean. And then I sigh. Pissed-off me has vanished almost as quickly as he arrived, and the nice guy is back in the driver’s seat again. I pet Bella’s ears and run my hand down her glossy coat. I scratch Bella’s rump and place my face near her mouth and invite Bella to lick my cheek.

Bella has cast me a life preserver. She’s nosed into my cheek with her cold, wet, loving snout. She leans into me harder, as if to gain as much surface-to-surface contact with me as possible. In this instant, I am calm. I am on dry land again, looking out at the raging storm over the sea as if it’s a disaster that happened to someone else. The storm has settled; the sky is clear and blue again. Bella looks up at me as if to say, "It's okay, Daddy, I'm here. Pet me. Remember that I love you. Remember that's all that matters."

My right arm reaches around her neck and I pull her in closer. I'm only now starting to realize how much petting Bella does for me. Whenever I am stressed, or anxious, or depressed, or lonely, contact with another living being is remarkably therapeutic. Bella has learned this about me. Sometimes I take the initiative and pet her. But often she senses my moods and comes to me first, ready to calm me down and be an anchor in an unstable world.

"I love you too, baby girl," I say.

My anger is still around, lurking under the surface. And maybe letting it all out isn't all bad. Far too often, I bottle up my aggression. I'm overly apologetic, passive on the exterior, trying so hard to spread positivity that I often fail to stand up for myself. But there needs to be a balance. I need to be assertive when needed, to open a door that allows stressors to flow out. I shouldn't close the hatch and bury all assertive feelings to the point that I boil over and blow my top. But I don't want to be out of control either.

And that's something I want to change

about myself. And it's something Bella's really been helping me with during our time together.

Thinking back, I could've used her help on several occasions before she was a part of my life. Most people know me as the nice guy, the guy who will do anything for anyone, anytime, anywhere. But those who know me the best, the ones who have been through the thickest of thick with me, know that if I reach that boiling point, what happens next can be uncontrollable. They also know that I always act on the defense. I've never gone out looking for a fight, and I've often been the one to mediate most of the situations I've come across. Yet if I'm provoked, if I feel threatened, if it comes down to fight or flight, I'll choose fight. Thankfully, I'm not very good at it, so I've never hurt anyone. I often joke that I've beat up a lot of dudes' fists with my face. I am lucky that my face can take a pretty good beating, because in these tussles I've never been knocked out, either. I'm not so good at landing punches, and I always take a few punches as I'm going in for the choke. I just want to let the person know that it's not okay to mess with me. That I'm not a sheep. That I'm not a pawn in their world of machismo where they can treat others as "less than." I really can't stand that. Yet that rage within me scares me a bit. It makes me question who I really am. Am I a nice guy, or am I out of control? Bella, of course, would offer a bit of a lesson for me in this.

We are hiking in a popular and heavily trafficked area where dogs are allowed off leash. The day is bright and sunny, the trail is full of a variety of people, and the dogs hiking alongside them vary even more. From small Chihuahuas to giant cane corsos, there is no breed that can't be found. Bella is ecstatic to see so many new friends to introduce herself to, especially now that she is off leash and has the freedom to say hello to whoever she wants. She meets a little Frenchie and they play for a bit, but the bulldog is on a mission to follow his owner in the opposite direction from Bella. A few people stop to pet Bella and admire her shiny coat that reflects the sun's golden rays. Bella accepts the compliments graciously and allows these new people to scratch behind her ears. Then she sees a group of three dogs and she's off! She runs right up to another Lab to say hello, but before she can, a shepherd mix pounces on her. Bella immediately rolls the dog onto its back and lets out a ferocious series of growling barks. She mouths the scruff of the shepherd's neck. The owner is screaming: "Get him off, get him off!" I call Bella, and she runs over to me, leaving the shepherd to collect himself. I lean down to see if she had any punctures, then look in the direction of the owner, who's still distraught from the altercation. "She's a girl, actually," I say with fatherly pride. "Check your dog for wounds," I say. "I'll bet ya he doesn't have any."

Sure enough, there are no wounds on the other dog. That's the beautiful thing about Bella. She always comes out on top during a scrap, but she never leaves a mark. I've come to call this "dogjitsu." Jujitsu is all about neutralizing an opponent, using their own force against them, rather than using your force to attack them. Good jujitsu practitioners will tell you that jujitsu is all about peace, about having the ability to resolve a physical conflict with the least violence possible. Though Bella can sound violent, she seems to use those sounds more as a warning. "Just so you know, I'm in control of this situation. I'd quit now if I were you."

The shepherd's owner gives me a scathing scowl and takes her dog in the other direction with his tail tucked between his legs.

"He started it, Daddy," Bella seems to say to me.

"Yeah, I know, Bella girl. You were just standing up for yourself. That'll teach him." And just like that, Bella is back to her happy self. She looks up at me with wide eyes and a wagging tail and then stays a bit closer to my side as we continue to hike up the trail. Perhaps she's learned a little bit too. What I don't quite realize yet is how much this teaches me about myself.

Bella and I make a good team because we are so much alike. We love people, we love new places, we love adventuring, and we both get distracted chasing squirrels. But it takes witnessing a few of these conflicts with her and other dogs before I realize that we are alike in another way. We choose fight over flight. We don't back down until we've proven to the opposing force that we are not pawns to be played with.

Bella, this extremely positive being, this dog that has done nothing but exude positive energy, this animal that has presented herself as the closest thing to perfection that I've ever witnessed, flips a switch and becomes an unrecognizable animal in order to protect herself. I know I can't fight like a dog every time someone threatens me, but as long as I learn to understand that the assertive part of me has its place, I can also understand that one part of me isn't all of who I am, because it sure as hell isn't who Bella is.

Bella and I are back on the road and we're heading toward Niagara Falls. I'm excited because the Falls are one of the most iconic sights in the country, and growing up, the only way to visit such a place was through a book or television. We park, and Niagara Falls in winter is everything you'd expect. Snowy and cold and powerful and wordless. Bella runs alongside me, hopping happily on her three legs and saying hello to everyone on the walkway, treating everyone she meets as a friend. I stare at my dog a moment and let the simplicity and magnitude of those words sink in. *Everyone is a friend.*

Bella can taste the mist in the air, and she's so joyful to see this giant rushing water spout, the sheer size of the place. I even shed a tear or two at the Falls, because it feels like Bella and I are truly taking this trip, side by side, experiencing things together that neither of us would have ever had the chance to see and do, if we didn't just . . . go.

As brilliant as this moment is, I still can't get my angry self out of my mind. *Who is this guy?* I wonder, because I know other young men whose rage flashes up, and I know I'm not the only one who wonders how to control his rage. I can see the man I want to be, and it's not him. The man I want to be isn't the guy with the flash temper. I want to be the guy who's got the confidence, skills, and presence to de-escalate the situation. Strong. Yes. A push-over. No. But out-of-control angry? No, not that guy — I don't want to be him. The guy I want to be like is my buddy Nick Peterson.

Nick, Avery, and I all met in boot camp and were instantly bonded forever once we learned we were in the same reserve unit in Nebraska. Nick was a great wrestler in high school and a strong athlete overall who could sprint through any obstacle course and hop up to a pull-up bar and crank out twenty like it's nothing. After the Marines, Nick became a pipefitter by trade, and youth wrestling coach by passion.

In 2014, Nick and I are at a Marine Corps birthday celebration and some active duty marine is hanging around our table. The conversation is cordial until he starts shitting on a reservist by insisting he's not a real marine. It's an argument as old as time, one I don't want to get involved in. Yet someone needs to step in and calm these boys down.

Nick looks at me, and I look at Nick, and I'm pretty sure what he's thinking, because when you're from Nebraska you don't run your mouth without somebody trying to shut it for you. The active duty marine keeps berating the reservist, and my hands are shaking, and the angry part of me is going to come out any moment, and I glance at Nick again. He's just sitting there drinking his Dr Pepper and chewing Copenhagen like a real country boy. The two guys stand up and move in close to each other. Nick takes another glance, stands up, walks close to the guys, and steps in between them. Both guys shut their mouths, and Nick puts his hand on the chest of the aggressor, separating them. I can't hear what Nick is saying because the room is so loud. But the active duty guy sort of nods and backs away. Just like that, the situation is over. My hands are still shaking, and I'm still wondering if I'm going to lose control, chase the guy down, and go ballistic. I'm seconds away from pummeling this guy's fists with my face, but I see Nick is already sitting down again, perfectly calm. He's a man who's so confident that he can both defend himself — and control himself. He can de-escalate a situation before it turns into a fight.

"And that, dear Bella," I say under my breath as we leave Niagara Falls behind us, "is the man I want to be."

We keep driving and stop by Lake Ontario, and even though it's chilly, Bella jumps in for a quick swim. She loves the water, and the sky is cold and blue. Bella runs out of the water, shakes, and introduces herself to a young golden retriever. Instantly they're friends. They scamper up and down the beach, and then both dogs meet an arthritic gray-faced chocolate Lab walking along with her owner. Another new friend. I talk with the owner and learn the Lab is fifteen years old. A part of me winces. Bella likely won't reach ten, let alone fifteen. I shake the sadness out of my head and replace it with purpose. I think, *We may not have much time left, Bella girl, but we are going to use it to live more than we ever have!* Bella brings me a stick, smooth from its time in the water: "That's the spirit, Daddy! No time like the present!" I toss the stick into the water, and she swims after it without a care in the world.

Over the next few days, we take several hikes, and after our long hikes, Bella always rests in the car. Her health is still remarkable, but running on three legs takes an incredible amount of energy, so I make sure she gets ample recovery time. I've learned it's a balance, making sure she gets enough exercise to stay healthy, yet not so much to injure her solitary front leg and shoulder. We travel through North Hudson, go see the Olympic arena in Lake Placid, hike in the Adirondacks, and the days tick by, and we keep driving. The clouds seem to lower; the air in front of us feels dense. It's almost Christmas, and I consider driving back to Nebraska quickly so I can be with my mother, but when I talk with her on the phone, she encourages me to keep going. She's been following my travels on social media, along with a small cadre of friends and followers I've connected with over the years, and Mom tells me she'd rather see us keep traveling than come home just yet. As much as Mom loves having me around, she has always been the one to tell me to go. She wants me to experience as much of the world as possible. In fact, so much of what I capture with my camera is because I want to share it with her. She tells me that my writing makes her feel like she's on the journey with me, which makes me smile. I call my dad as well: "Keep going," he says. "You're doing all the things I wished I could've done at your age." I know Bella and I are blessed to have this freedom, and I share our journey so that others can take it with us.

On Highway 22, Bella and I see a woman dressed in nearly all black. A stranger walking beside the road. Her hands and face are windburned, like she spends much time outdoors. She's passed middle age and wears a long black overcoat and hiking boots with a purple knitted scarf wrapped around her neck, all topped with a knitted stocking cap that allows the back of her hair to bunch out freely. She's a bit of an oddity, and we wonder what she's doing, walking along the highway like this, but almost as soon as we see her, she's already gone. "Hey, where'd she go?!" Bella seems to say.

Not far down the road, we stop at Port Henry and snap a photo of a captivating old church that feels like it belongs on the set of a vampire movie, but in the most romantic way. The church was built in 1887 and its paint is chipping off the wood, but its bright red front door still holds its color. An elderly woman who lives next door comes out, and she tells me she's the last active member of the congregation and has tried everything she could think of to keep the church going, but now it's dead. The church didn't adapt. A man bought the church building and today uses it as storage for junk. I make a mental note. So many things in life die when they simply refuse to change.

We say our goodbyes, and Bella sniffs around the old church carefully. She stops quickly and stares intensely into the nearby woods, quavering as if someone is there but not seen. Maybe it's just a squirrel. We climb back in Ruthie and start driving out of town, but before we get far we stop at an old train station that was built in 1888 and still operational. I view the station through the lens of my camera and snap a photo.

Next to me a voice suddenly says, "How do you like our train station?"

It's the lady in black — the stranger we saw walking beside the highway. Her sudden presence spooks me, but Bella leans in for a friendly sniff and I give the woman a closer glance. She looks sane and isn't wielding an ax, so we talk and I tell her I'm on a journey with my dog, and she bends down to pet Bella and then asks, "Would you both like to take a walk with me?"

Her request feels a bit strange. We don't know each other. But maybe this is the advantage of being a man traveling solo, particularly with a dog. If our genders had been reversed, and a strange older man dressed all in black asked a young woman traveling alone to go for a walk, I'd shout: *Halt! Wait! Stop! Scary movie ahead.* Overall, I'm intrigued and say I'd love to join, and she and I and Bella amble across the railroad tracks and into the woods. I've made the judgment that we aren't in danger, and something tells me that Bella's smiling face and joyful spirit are what allow the woman to feel the same.

The woman leads me down an unbeaten path strewn with leaves and foliage, and I wonder if the path we're on was created by her. The more we talk and walk it's clear she simply wants to show us her backyard, a wilderness she's proud of. No scary movie here, rather another great human connection. Her path leads us through a boatyard with shipping containers and old boats on stilts, and we reach the shore of Lake Champlain and stop. She notices I'm taking photos all along our walk and says, "Okay. That's enough pictures. Put away your camera and enjoy the moment."

It's a fresh reminder, something I need to hear, as I've collected more photos already than I know what to do with. My head is always so full; taking photos is the best way I can remember things, fleeting moments in time captured forever in a single image. But

I set aside my camera for a while and stare at the lake.

She tells me how she used to own a home and have a job where all she ever did was work. But she decided one day that the life she wanted wouldn't be spent working every moment until she died. So she sold her house, quit her job, and started going for walks every day.

I'm curious how she lives without a job, but out of politeness I don't ask. People ask the same about Bella and me, saying they wish they could just travel and have no responsibilities. The moments I share on social media are mere highlights that even I am envious of, but it's far from the whole story. Our journey hasn't come without sacrifice. Bella and I have shared many lonely nights when I wonder what it is we're doing, flooded with anxiety that perhaps this trip is only setting me back further from life goals. Thoughts of owning a home, maybe some land, and possibly having a family of my own someday loom in the back of my mind. I know I can't afford to do that by selling a few photos. Yet it's enough to allow us to live freely now, and when will I ever have the chance to do this again? With Bella, never. This is our only chance. With myself, I've seen friends and loved ones die before they ever get the opportunity, working their hands to the bone, saving for a retirement they never lived long enough to enjoy. When you see the dreams of someone close to you die along with them, it's hard not to chase your own dreams if the opportunity arises, *now,* before uncontrollable circumstances steal it from you. Right now, we have that opportunity. We have that freedom. Freedom to take the gravel road to nowhere. Freedom to walk with a stranger and enjoy the moment.

Our conversation is over, our walk is done, and the lady in black says goodbye with a smile and disappears into the woods. I'm sure she has told me her name, but I can't remember it. I stare after her, thinking about what her life experiences will continue to be as the woman who walks the lone path along the railroad tracks. I'm curious how many Robs and Bellas she will invite to walk that private path. We will probably never see her again. But I'm hoping that Bella will draw other people like her toward us as we travel, strangers who become friends.

Bella and I drive into New Hampshire, and a military friend named Patrick reaches out through Facebook and invites us to stay at his chalet. He's retired now, just him and his dog, Cruzzah, and we spend Christmas with Patrick. He takes us over to his friend's house, and we sing karaoke, and they love on Bella, and we're blessed with warmth and acceptance in the houses of Patrick and his friends. Holidays have always been tough for me, because of the broken-family dynamic, but now I use them as another excuse to spoil Bella. A new chew toy is the standard, along with a few treats. This year, a bully stick is her present. It's not until after she's finished it that I realize this may be our last Christmas together. Suddenly a measly bully stick doesn't seem nearly enough.

The morning after Christmas, Patrick leads us on a hike through the snow to an old bridge. Bella burrows deep through the snow. It's to her chest, and she's loving it, and I never want this moment to end. I see the smile that grows on Patrick's face as he watches Bella joyfully plow through the snow. I've seen that smile before. The lady in black. I haven't realized it until now, but it dawns on me what a smiler she was. She smiled at the lake, and she smiled for most of our walk. She had smiled when we first met her, and she smiled when she petted Bella — a really big smile where she showed all her teeth, a smile that took all her face, and one where the smile lines formed in her eyes when she squinted — and she smiled as she told me to put away my camera and live the moment rather than capture the moment. She was smiling and smiling, and I wonder out here in the snow why I hadn't noticed this smile before. Come to think of it, she'd been laughing too. A gentle laugh, a pure and fun laugh that's always in the background of the conversation, a laugh full of a wisdom I am only beginning to glimpse.

I consider how my short interaction with this woman and with friends such as Patrick and Cruzzah, and the time I have spent with Bella and the open road have all combined to feel more therapeutic than anything I've experienced in a long while — and I guess Mom noticed this in my voice, in my photos, and could see that this trip is great for Bella and really good for me too. Bella smiles all the time — it's part of why I love her. Suddenly I notice I am smiling too. The old church in Port Henry might have died, but I won't. It is possible to change, to adapt, to grow, to mature. I've watched Bella do it since the amputation; she's overcome the physical hurdles and hasn't let the emotional impact touch the sides. She's adapted beautifully to her new circumstances. She hasn't been angry and she hasn't been depressed.

I reach down and pet Bella. The uncontrollable me has begun to fade away. That anger will only come out when needed, and if and when it does, that part of my personality is under my charge, not the other way around. I am becoming the man I want to be.

I like that idea.

I like that idea a lot.

Bella gives me a big smile. She likes it too.

5
Gentle Warrior

It seems like so many years ago now. I'm mowing the lush green grass in the front yard of our home. Bella sits on the porch, off leash, smiling in the sun. She's fully grown in size, but a puppy still at heart. Every so often she gets up and tries to creep out into the yard, but I catch her, shut off the mower, and say, "Nuh-ah. Back on the porch!" She tiptoes back up and lies back down with a sigh of discontent.

Our letter carrier is new. He comes by, scowling. I can tell he isn't too thrilled about her being off leash. I shut off the mower and call out, "She's a special girl, man. She wouldn't hurt you if I told her to. Come and meet her, you'll see."

He shakes his head. "Nope. Been told that before. Been bit before."

I walk up to where he is, and he hands me the letters and bills and continues on his route. I take the mail to Bella, but she doesn't seem interested in taking it from me to walk it the three extra steps inside. I chuckle as I bend down to give Bella some good tummy rubbings. "You aren't a biter, are ya. We'll turn him around. We'll show him not all dogs hate mailmen."

A few weeks later we're packing the car for an afternoon out, when I hear: "Is that your dog?! IS. THAT. YOUR. DOG!?"

I spin toward the sound, and Bella is trotting happily over to the same letter carrier, who is walking through our grass in the direction of the mailbox, which sits on the side of the house. "Yeah, she's nice!" I call out. "Hold on, I'm com—" Before I've taken even two steps, he sprays her in the face with a canister in his hand.

Bella stops fast in her tracks, winces.

Fatherly instincts kick in, and immediately I'm over the top. I sprint over and grab Bella by the collar, check to see if she's okay, and lay into the mailman. "Are you kidding me? How many times have I told you she's nice?! She's never had a single reason not to trust a human, and you just gave her one! Get out of here!"

"She shouldn't be off leash," he replies calmly. He hands me my mail and continues down the street.

She's in my own yard, I think. My blood is

boiling and I want to tackle this man to the ground and beat some sense into him, but instead I inspect Bella more closely and see she doesn't appear to have any lasting effects. Her eyes aren't bloodshot, she's not sneezing, and she seems a bit more confused than anything. In my younger days, I was hit with pepper spray once. It was awful. I was blinded for minutes, and my entire face burned for two days. Bella definitely didn't just get a snout full of pepper spray. Perhaps citronella? I've seen those canisters at pet shops. They're recommended for dog owners to carry while walking their own dog, in case they need to ward off an aggressive dog that's off leash.

Damn. *Off leash.* It hits me. As good a dog as Bella is, the letter carrier was just doing what he thought he needed to do to keep himself safe. I remember how he'd mentioned in our earlier conversation that he'd been bitten by a dog whose owners told him not to worry. Experience had driven his self-protective behavior, and I couldn't get mad at him for that. I'm the one who had Bella off leash when he needed to come up to the house. I'd have done the same in his situation. I'm the one who needs to say sorry.

I put Bella inside, jog down to the end of

the block, and catch up with the letter carrier. I apologize to him for losing my cool, as well as having Bella off leash while he was doing his rounds. He tells me I could be written up for having a "dog at large," but he appreciates my apology and won't be turning in any paperwork. I have to stifle a private snicker at the thought of Bella being labeled a "dog at large," as if she's a vicious beast, eager to hunt small children. But I bite my tongue and thank the letter carrier for not turning us in to the authorities. Bella's record is squeaky-clean, and I'd hate for her to have a conviction for something so trivial. We shake hands and I head home, encouraged at my ability to quickly forgive the guy and somewhat annoyed that he still wouldn't give Bella a chance.

When I walk in the door, Bella greets me with a wagging tail. She appears unaffected by the incident, and I explain to my girlfriend what happened.

"Well, babe . . ." she says, "even though you and Bella want to be friends with the entire world, someday you'll learn that not everyone wants to be friends with you."

If only all the world could be friends. Back when I was fifteen, I watched Mike walk across the parade deck at the Marine Corps

Recruit Depot in San Diego as part of his graduation. The memory was seared so brightly in my mind. I thought my chest would burst with pride. There was no doubt that I would follow in his footsteps someday and become a marine myself. I'd observed the monumental changes that occurred in his life for his own good, and I saw the respect he gained from others around him. I saw the focus of his mission, and I knew I wanted those things in my own life. The Marines offered a chance for me not only to serve my country, but to become a part of something bigger. A chance to serve. A chance to belong.

My brother was tremendously proud to be a marine, but he also strongly encouraged me in a specific direction. Back in 1998, it wasn't what I expected to hear.

"Bro," he said one afternoon when I was sixteen, still in high school, and preparing my paperwork to be eligible to enlist, "don't go active duty. All we do is run our dicks into the dirt and break our bodies for no reason. It's not like we're going to go to war anytime soon, either. Everybody has nukes. Go reserve, man. You'll still be a marine and can go to school or use whatever job they train you for in the civilian world. When us infantry guys get out, we don't have many options. In the infantry, guys are leading platoons when they're barely over twenty years old. But when infantrymen get out, they're working security jobs for nine bucks an hour."

It made sense, so I followed Mike's direction. I admired Mike, and I imagined the Marines offering me the chance to become part of something bigger and better. I wasn't looking for a fight, but if our country ever needed to go back to war, I wanted to be in the fight. I didn't want to be stuck on the sidelines again. Mike had been in the Marines for three years by then, becoming a sergeant in only two and a half. He was a primary marksmanship instructor and a jungle warfare instructor. The dude was hard core. He was well respected by everyone, even used as an example by recruiters. When Mike told me to go reserve, I obeyed.

There are countless jobs to choose from in the Marines, from tank crewman to aerial gunner on a helicopter. No matter the job, I wanted to be one of the few and the proud. When I spoke to one of the recruiters at Grand Island, he assured me that "Every marine is a rifleman." The line means that all marines are marines, period. If you're a cook in the marines you still carry a rifle, you still know how to shoot, you still get the job done taking down the bad guys. The recruiter phoned up what would later become my unit in Omaha and asked, "What do you guys need down there?" Nebraska had one reserve unit, and it was a maintenance company. The recruiter set down the phone and added: "Okay, Kugler, you're gonna be a 1341."

"What's a 1341?" I asked.

"Heavy equipment mechanic. You're good to go."

"But sir," I asked with a sixteen-year-old's sincerity, "if we ever go to war, I won't still be a mechanic, will I?"

He stood, put his hand on my shoulder, and said, "Son, every marine is a rifleman. If your country calls you to war, you'll go to war for your country."

The day I turned seventeen I drove down to Omaha and officially enlisted. Mom had already signed the waiver so I could join before turning eighteen. It was a sweltering summer day, four weeks before the start of my senior year of high school, and the Marines were more than happy to accept me, although I wouldn't actually head off to boot camp until I graduated from high school the following June.

That fall, when I started my senior year of high school, I had no doubt the Marines were in my future, which kept me out of trouble while I watched several close friends fall into drugs and drop out of school. The Marines had a zero-tolerance drug policy, which kept me from following suit. I still didn't know how the various pieces of my life would all fit together. Quietly I held dreams of becoming an actor someday too. To those who knew me best, I was a performer, a comedian of sorts. I'd impersonate famous characters and even mimic people in my life. Early in my senior year — in the knowledge of being funded by the reserves — I looked at the idea of majoring in theater in college. I talked to a few teachers about my plan, but none seemed excited about it. One teacher scoffed and said: "Let's be honest, Robert, what's ever going to happen with that? Don't you plan on joining the Marines? You should just stick with that."

The seventeen-year-old me took that comment to heart. *Okay.* If nobody else thought I could do it, then I wouldn't believe I could do it myself. That seemed to be a pattern with any educator I talked to. I don't blame them; they were trying to look out for what was best for me, a tried-and-tested route for a kid like me. A few weeks later, when my advisor asked me if I was going to go to college, I told her, "Nah, I'm joining the Marines," to which she quickly replied "Okay" and snapped shut my file.

That settled it. I knew the Marines' core values were honor, courage, and commitment, and I wanted what the Marines were offering. I was going to pursue that pathway of life with everything I had.

All Marine recruits who live west of the Mississippi River are shipped off to San Diego for boot camp. If you live east of the Mississippi, you're shipped to Marine Corps Recruit Depot, Parris Island, in South Carolina. If you have a college degree and are planning to be an officer, you're shipped to Quantico in Virginia. I went to San Diego.

It's not easy to become a marine. Out of all branches of the military, it's said to be the most intensive. For thirteen weeks, I and all the rest of the new recruits rose at 0600 hours, ran and drilled, drilled and ran. Our drill instructors shouted at us, insulted our mothers, called us jackwagons, and only allowed us to speak after we'd asked their permission — and even then only in third person, all in the famed effort to tear us apart and put us back together again in their image. You are not an individual. You are not your name. You are a recruit. You are government property.

We learned the basics of hand-to-hand combat, chanting: "Slash, Buttstroke, kill!" as we attacked the air with dummy rifles. On the rifle range, we learned to shoot from five hundred yards with open sights. The seeds of the warrior spirit were planted deep within our souls. The history and traditions of the Marine Corps were ingrained into our minds. This newfound comradery was branded on our hearts.

Three weeks before graduation I received some difficult news. Mike's unit had been called away to fight forest fires near Salmon, Idaho. He couldn't make it to my graduation. For a moment, that broke me, sending tears down my clenched jaw as I read the letter. The image that constantly motivated me through the hardships of boot camp was Mike congratulating me on the other end, wearing his dress blues, smiling and proud. I wanted other people to see him next to me. Mike the sergeant. Mike my brother. I couldn't wait to show him off. My dad couldn't come to graduation either, but my mom did, along with my sister Amy and her two-year-old son, Chandler, my nephew. We took a picture with Chandler on my shoulders at graduation, which is still on Mom's refrigerator today.

I felt proud — actually, I felt *incredible* — to call myself a United States Marine. To be given the Marine Corps emblem, the eagle, globe, and anchor. In the army, you're handed the rank of private even before you complete boot camp. But in the Marines, you're a "recruit" until you complete the Crucible, a grueling fifty-four-hour test that must be completed before graduation. After forty-five miles of hiking, the last trial in the test is hiking up a mountain called the Reaper. Our legs burned hotter than the desert sun that shone down upon us as we conquered its steep grade. Just as I felt I couldn't make it another step, one of my buddies started falling back. His feet were blistered raw, and he said through exhausted breathing, "I can't make it!" I ran over to him, grabbed him by the pack on his back, dug down deep, and literally started to push him toward the top. When I was too exhausted to continue, another recruit took over for me. We all made it to the top together. It was one of my first true lessons in the power of service. Of selflessness. I had almost given up on myself, but when I helped someone else, I was stronger than I've ever been.

I'd find that same quality in Bella years later. I'd see her resiliency and draw from it. My mind wouldn't have time to wander to places of self-doubt or pity. I'd pay attention to her to ensure she stayed safe and healthy, to keep pushing to inspire her to live, to simply be there for her. You learn that when life isn't all about yourself, it's actually much easier to live.

There, in Marine training, once at the top of the Reaper, an eagle, globe, and anchor was placed into each of our hands by our drill instructors. I closed my fist tightly around this sacred emblem, this tangible rite of passage. The marine next to me said, "Kugler, we did it, man. We actually did it!" He was a stocky, stoic fella, and tears streamed down his face. Tears streamed down mine too. In that moment, I became more than the poor kid from a broken home. I became a *marine.*

Boot camp was only the beginning. After boot camp came Marine Combat Training, three weeks of sleeping no more than an hour each night, humping our asses all over the hills of Camp Pendleton while learning how to be better riflemen. I ended up being the radioman for our patrols, which meant I carried my pack in addition to our platoon's radio, the extra weight made it hard to keep up, and more than once I felt like I wanted to give up. But just when I was about to pass the radio off, a marine said he was motivated by me. He saw how I was going so hard and not complaining. He had no idea how close I was to quitting. So I kept my mouth shut and trudged on. This taught me the power of encouragement. A few simple words from someone else, at exactly the right time, restored my faith in myself. That faith somehow translated to physical energy. I've never forgotten that power, and ever since I do my best to encourage others. Every so often on a hike, Bella will encounter some obstacle and pause and look at me, maybe wondering if she can cross a stream, or traverse a fallen log in the path. "You got this!" I'll shout, and slap my knees. "You got this, girl!" With that simple bit of encouragement, she's able to tackle most any obstacle. Only when she absolutely needs help do I step in.

There's just something about creating bonds within a community that really makes it feel like somewhere you belong. Whether that's the guys I trained with or the people who live around me, I've always been quick to make friends, to create and embrace community. Knowing your neighbors builds trust and connection within your immediate surroundings, and when you become friends with them, the world simply seems to be a safer and happier place. Neighbors become friends, and friends become family.

Bella is fully grown but still young, and my girlfriend and I have met our neighbors, who are grandparents. Their daughter, Francine, and a young man named Julio have a young son together, Adam, who's not quite three years old. The grandparents ask if Adam can meet Bella next time he comes around. I say sure. Next day, Julio pulls into the neighboring driveway and gets Adam out of his car seat. I walk over with Bella beside me, and Julio turns to the side, instinctively shielding Adam from any potential harm.

"Don't worry, bro," I say. "She's amazing with kids. You can trust her."

My girlfriend and I always planned on having children while we had Bella. Although Labs are notoriously gentle natured, we still took some steps early on to ensure she'd be able to tolerate any unknowing little human attacking her. Ever since she was a puppy, I would get down and put my face in Bella's food dish. I'd take away her toys. I'd pull on her ears. I'd pinch her skin and pull on her tail. Basically, I'd do anything a little kid would do, because we wanted Bella to be unfazed by this type of behavior.

Julio sets Adam on the ground, and Bella walks over with cautious eyes. She approaches in a way that shows she isn't a threat. Her head is lowered, her tail wags softer, and she takes considerate steps, making sure not to knock him over. Gently, she raises her head to sniff Adam's face, then softly licks the remnants of a recent snack off his cheek. The toddler laughs and takes a step back at the same time, not so sure what to think of this wet-tongued furry creature lapping up dried milk and cookies from his face.

Bella had always been curious about kids. If she heard a baby crying down the street, Bella would run to the screen door and tilt her head from left to right, curious as to where and why a child was in distress. Whenever our friends came over to visit with young babies in their arms, Bella would jump up to try to see them. "C'mon, Daddy, I just wanna see the baby! You're all passing it around, why can't I see it too?" Normally, Bella was a good listener, yet she's been deaf to our words telling her to stay down, so we learned to ask our friends to bend down and let Bella sniff their baby, and then she would be fine. Our trust in Bella was built over time, an absolute and wholehearted trust, a trust that I love sharing with other people.

Bella takes another step forward. Adam reaches toward her, grabs one of Bella's ears, and tugs. In Adam's other hand, he holds an uneaten peach. Instinctively, he gives it a toss. Bella runs over, picks it up in her mouth, and walks slowly back to Adam. She sets the peach down delicately at his feet. Before Adam can reach for it and take a bite, Julio picks it up and exclaims, "Wow, there's not even a tooth mark on this." I watch a smile grow on his face as he looks to me with bewilderment and adds, "That's awesome, man. Did you have her trained or something?"

I explain how we've worked with Bella since she was a pup, but it's also her breed and temperament. Labs are known for their soft mouths. Bella's a gentle, curious pup by nature.

"She's just a sweet girl, man," I say. "That's just who she is."

I'm not sure if Bella has ever understood this power of hers to bring people together, to make you feel good, but she sure as hell has used it time and time again. She'll come in for contact. She'll go grab a toy, perhaps her "boingy-boingy," a rubber spring shaped like a DNA strand, and shake it vigorously.

She'll start spinning like a bucking bronco with it, and suddenly life grows a whole lot brighter.

So many years later, with Bella on three legs, the rodeo action has become nearly impossible to do, but she finds other ways to encourage me to lighten up. Maybe bringing me a sock to play keep-away with it. "Hey, Daddy! Did you know you left this perfectly dirty sock in the hamper?! Here, I brought it out for you to play with! No . . . no, you can't have it! Try and get it! Come catch me!" Or she'll growl as if she's the most vicious warrior dog on the planet, and I'll get down on the floor with her and play tug-of-war with the sock in my teeth, and she'll be gentle enough not to rip it from me. If I ever give her the command to drop it, she'll release it immediately. Then I'll gloat as the victor of the sock war, until I realize my prize is a dirty sock in my mouth.

After MCT, I was sent to Missouri for my military occupational specialty training, where I'd learn the intricacies of becoming a heavy-duty mechanic for the Marine Corps. All the equipment and supplies shipped overseas to war zones or humanitarian missions are moved by heavy equipment. Marine engineers build military bases with heavy equipment. Machines break down under the strain. They get clogged with sand. Teeth get chipped, and rotors warp, and engines sputter and stall. Marines break it, so marines fix it. My job wasn't a frontline combat job, but I was a marine first, a rifleman first. If our country needed me to go to war, I would go.

I graduated second in my class from MOS school and was awarded the "Gung Ho" award by my peers for being the most motivated. We only had a class of 6, so I can't brag too much about that. My class commander, a gunnery sergeant, was brokenhearted when he found out I was going into the reserves. "We need marines like you in the fleet!" he said. It felt like the biggest compliment of my life. I'd found my place in the Marines — and I liked it.

Yet the feeling was so momentary. In an instant I was back home after graduation, back in Nebraska as a marine reservist. Still a marine, but not feeling like one. Not surrounded by the team anymore. The pack. The brotherhood. The purpose. Now what? I was eighteen years old and had just gone through an incredible transformation. Being a marine had become my new identity. As a marine, I belonged. But the reality of being a reservist set in: Drill is only one weekend a month. Who was I the other twenty-eight days? And what sort of purpose would I find if I wasn't wearing my uniform?

■ ■ ■ ■

Two: A Long Way from Home

■ ■ ■ ■

6
Lemons

I'd become a preschool teacher! It was the end of the summer, 2001, I was still a marine reservist, taking college courses full-time. The GI Bill at the time was only $272 a month, hardly the "will pay for college" idea I had going in. So, I had to find full-time work to pay the rest of tuition and for my living expenses. Being the youngest of seven, I grew up with ten nieces and nephews, so taking care of kids had been a part of my life since I was a kid myself. It only made sense to put those skills to work. I started out working with teens, then found myself working with the younger kiddos. Days, I worked at a preschool, doing something I felt mattered in the world — helping the next generation along their way. I loved my job with the children. I'd obtained my Child Development Associate certificate, and a number of the kids in our school had come to us because we had a "no child is turned away" policy. This didn't mean they were bad kids, it meant they needed help, and the teachers at our center lived to give these kids the best chance possible. The more I worked with these young souls, the more I often glimpsed my younger self in their faces. Whenever I consoled a child who insisted, through crying eyes and a runny nose, that "no one cared," I hugged him and said it wasn't true. *We cared.* If I closed my eyes, I was hugging five-year-old me.

September dawned bright, although the brightness held a curious tension. A friend called me early morning on a Tuesday that same month and said I needed to turn on the news — quick! What I saw on TV was a smoking tower on the skyline of New York City. Reports were hazy at first. All anyone knew initially was that a plane had collided with the North Tower of the World Trade Center. *What the hell? Surely a plane could have avoided such a large building,* I thought. In bewilderment, I watched the television — live — as seemingly out of nowhere a second plane came into view and struck the South Tower. My heart dropped. This was no accident. My brother's "We'll never go to war" comment ran through my mind again.

The first tower collapsed and fell. The

second tower collapsed and fell. News came that the Pentagon had been hit. Flight 93 had gone down in Pennsylvania. Sure enough, the news broke that all the day's horror was the result of a massive terrorist attack against the United States. Innocent blood had been spilled on our soil, and I envisioned how our country might respond. A fire began to burn inside me as I pictured deploying with my marines to fight the terrorists who'd just attacked and killed citizens of the country I called home, the country I'd sworn to protect. How will we respond? It sunk in. I was a part of that "we." Attacks such as these were why I joined the Marines above all other branches. Because "Every marine is a rifleman," and when my country needed me to go to war, I would go to war.

Minutes after the news reported terrorists were to blame, my phone rang. It was my squad leader from my Marine unit. He said simply: "Get ready, Kugler. Pack your shit." Then hung up.

I was, and I did.

But no further call came.

It's a funny feeling, wanting so desperately to do your part, but being told no, you can't. There wasn't a rush for reservist marine mechanics to join the invading forces that hit Afghanistan, and for the next two years our unit was placed on standby, anxiously awaiting our call with our go bags packed.

It wouldn't be until 2003, when America entered Iraq, that our unit would get the chance to be part of the global war on terror. In 2003, my Marine and National Guard buds gathered at my apartment to watch the invasion of Iraq. My TV screen was dark and green, the images viewed from the perspective of a night-vision camera, as tracers lit up the sky on their launch route into the city of Baghdad. "Shock and Awe," it was called. The fury and intensity in the room was palpable. Our unit had missed the fighting in Afghanistan, but surely we would be called up for the invasion of Iraq.

Our call wouldn't come like we imagined it would. Unlike other branches, our unit wasn't set to deploy all together as a whole. Instead, the higher-ups took a handful of reservists from one unit, and a handful from another unit, and so on and so on, and pasted them together to form a hodgepodge of a new company. This went on several times, and every chance I got, I raised my hand and volunteered to go, but time and time again, my name was passed over.

The main reason I'd missed my chance

was because I'd volunteered too early. Once I'd realized we didn't have a chance to deploy to Afghanistan after 9/11, I'd volunteered for a nine-month Marine humanitarian mission to build roads in Central and South America. I'd even pulled out of college for it. The mission to South America ended up not happening, but my name never made it back onto the list for members of our unit to be deployed to Iraq, despite my pleas to my commanding officer.

One by one, I watched my friends deploy as I stayed home. My brother Mike got deployed, while I was still teaching preschool. Stateside, I was promoted to corporal, and then to sergeant. Mike came home to pin on my chevrons, along with my mom. Our country was at war, my friends were at war, my brother was at war. They left, did their part, then came back again. I simply wanted to do mine. I made major adjustments in my life — packing in my preschool job and taking odd jobs to pay the bills, quitting college, giving up leases — to ensure that I was ready to leave the moment the call came. A couple of times I made it onto deployment lists, only for it to fall through. Twice, departure dates came and went. I can't explain how much of a mind-warp it is to be told you're going to war and then not go. I was stuck on the damn sidelines.

While waiting, I struggled to find the right direction forward, often partying and chasing girls. I went back to community college and switched my major to health and human services, and then switched again to fire protection technology. All the while, working full-time and still in the Marine Reserve, not only waiting my turn to deploy to Iraq, but chomping at the bit to get my chance.

Some things in life are like that. The situation is rough, and you just have to deal with it. Bella reminded me of that truth more than once. She might have once picked up a peach as gingerly as a mama cat picks up a kitten, but not all fruit that went into her mouth received the same treatment.

Once I brought home a few lemons and set them on the kitchen counter. She was maybe two or three years old then. Still very much young at heart. A lemon rolled off the counter and Bella ran to investigate. She'd never seen one up close. Before I could tell her no, she scooped it up and squeezed it like a chew toy, then spit it back out and shook her head, unable to take the tartness. She picked it up again, then immediately spit it out. She backed up and pounced her feet in front of it, barking wildly as if she could scold the lemon into submission. "Why do you do weird things to my mouth? Just be a regular ball!"

Normally, I'd take something like that away from her, but I was having too much fun watching the situation unfold. I opened the back door and held it ajar. She walked around the lemon, eyeing it as if it were going to jump at her, then scampered outside and ran in circles.

"What, are you scared of this thing now?" I asked as I scooped it up and tossed the lemon in her direction. She loped over to it, pounced on it with her front feet, then ran away. She came back to it, sniffed it, picked it up with her mouth, and flung it high in the air. The lemon landed on the ground behind her, startling Bella. She took off running to the other side of the yard, then whipped around and darted back to the curious citrus, again reprimanding it for its existence with a series of barks.

This went on for what seemed like an hour. Finally, Bella submitted and lay down by the fruit, panting as she stared at it with puzzled eyes.

"Well, Bella," I said. "Some things just are

the way they are. We have to learn it's not our place to change them."

How I wish I had learned that lesson earlier. Or learned it *better,* perhaps. As I waited and waited for deployment, someone wholehearted and noble came into my life. In time, I would try to change this person, not having learned yet the lesson of Bella and the lemon.

One of the many jobs I worked during this season was as a doorman at Christo's Pub in downtown Lincoln. Christo's was two blocks away from the main strip of college bars, and the patrons were a bit older and more refined. On Husker game days, every business with an open door was home for the sea of red that poured into downtown to root for our team. Husker football is a second religion in Nebraska.

At Christo's, I had the dual mission of checking IDs and slinging burgers and bratwurst in front of the restaurant. I cooked on a grill right on the busy sidewalk, and as I cooked I hollered out in my best Scottish accent: "Aye, we got yer burgers and bratwurst here for ya foooolks. Come an' get it before the big foo-oo-oot-ball match!" Not sure why I went with Scottish. It just felt great to belt out my pitch with that accent.

One evening, two young women stepped up to order and inquired if I needed to watch *Braveheart* before each game to get into character, which made us all chuckle. This started a conversation that ended up with introductions. One woman's name was Charli. She had fair skin and beautiful chestnut hair and mostly brown eyes, ringed with green around the outside.

Charli.

That was her full given name, and it wasn't short for Charlene or Charlotte. I was instantly intrigued. The women went back inside the restaurant before any numbers were exchanged, but later that evening I discovered that Charli was a good friend of a marine in my unit.

Over the next weeks, Charli and I bumped into each other a few times downtown and said hello. At Sandy's pub, we finally connected and I asked her out. We dated casually at first. She was educated, worked as an accountant, and already owned her own home at twenty-three. Her life was in order. I warned her my life was basically on hold. I was living with a buddy because of the last failed deployment that had left me near homeless. I also told her that I wasn't looking for anything serious, that I was a rolling stone with my sights set on Iraq first and foremost, and if I didn't go there soon, I was headed to California. Maybe it was finally time to pursue the acting dream. But nothing about my lack of certainty seemed to faze Charli.

I went solo to the 2006 Marine Corps ball, because I was set to be the master of ceremonies of the event, and dates never have a good time sitting by themselves. When Charli found out I was going, she went with Kyle, a mutual Marines friend. After I finished my hosting duties, Charli came over and sat with me. I was captivated. From that evening on, we were a couple.

Back home, the buddy I was living with joined the army and moved out of his house. I needed to find a new place to stay. Friends in Dallas invited me to move in with them, and I considered moving there so I could take voice-over classes. But then it was Valentine's Day, and Charli gave me a card. Inside was a key to her house.

"You want to move in with me?" she said simply.

I couldn't help but smile. In Charli's world, things were just simpler, easier. I needed more of that in my life.

I moved into Charli's house, and in this new role together we clicked instantly. We just fit. And the relationship worked in big ways. Her family was supportive, and I experienced the beginnings of a new, secure world. I wanted to make this relationship work too. Nights at the bar making bad decisions turned into nights on the couch, watching movies. A search for purpose transformed into a feeling of calm and contentment. Charli was the best thing that had happened to me in a long while.

The home Charli invited me into was perfect. This was the first serious relationship I'd had in years, and we got along really well and were pretty sure we wanted to stay with each other, even start a family someday together. It became our home together. I was the happiest I'd ever been, and perhaps the most stable. Maybe those two things have something in common. I was never happier spending time with one person than I was with her.

Finally the call came. I was going to Iraq. But my eight years of service were set to be finished mid-deployment. That meant the only way I could deploy was if I reenlisted. Did I truly want this?

Yes. I wanted this more than anything. I simply could not conclude my time in the Marines without first being deployed. This lack of deploying had felt for a long time like a gaping hole in my heart. Our country was at war, and this was my duty, my sense of service, my purpose. I signed up for another three years in the Marines, no monetary bonus, only the opportunity to deploy. Finally, I would be able to fill that void.

That's when we found Bella, so Charli wouldn't be alone when I was overseas. It would be a twelve-month mission total: five months spent in pre-deployment training at Marine Corps Base Quantico in Virginia and Camp Lejeune, North Carolina, then seven months at Al Taqaddum Air Base in Iraq, just west of Fallujah.

In pre-deployment training, we woke early most mornings for physical training and ran anywhere from three to five miles. We did lunges across football fields. When we went to the maintenance shop, rather than take the time to brush up on our job skills, we inventoried the tool room and stood around looking at pieces of equipment so lightly used they needed no repairs. We sat through countless classes on the rules of engagement and escalation of force (we called these classes "Death by PowerPoint"). We updated our wills and made sure we had our power of attorney set up. In the field, we learned convoy tactics and trained in combat scenarios, patrolling through the woods and attacking imaginary enemies in makeshift towns. I liked the chaos and excitement of the training exercises. I felt I was in my element. And, importantly, there was no time to be in my own mind. The depressive me, the me without purpose or vision, all but melted away. All that mattered was each moment I was in. My expectations for deployment were high. Engaging in combat overseas seemed a definite possibility. I hoped above all that I wouldn't have to spend all my time overseas as a mechanic, because I sure as hell wasn't training like one.

Bella was just a puppy, only a few months old, when I finally left for Iraq. Already I'd been taking her everywhere. Already she was my little shadow. On the morning I left, I hefted the first load of gear out to the back patio and returned for another load. When I came back out, Bella was curled up in a tiny ball on top of my sea bag (the long green military-issue duffel bag). She gave me that inquisitive puppy look as if asking: "I'm coming on this adventure with you, right, Daddy?"

"Aw, baby girl. Not this time," I said.

"You're going to stay home with Mama, and Daddy's not going to be back for a very long time. But that doesn't mean he will ever stop thinking about you."

Her little brow furrowed. I bent down and petted her glossy coat. After seeing how Bella and Charli got along so well with each other, I was certain our decision to get a puppy was a good one. Deployments are hard on the service members who deploy, but they are also hard on our loved ones who stay home. Hopefully the companionship that these two offered each other would help ease that sorrow.

Already Charli had proved an amazing mom to Bella and had worked hard to make sure Bella was a well-mannered pup. Having a dog that begs or doesn't listen wasn't an option for Charli, and she had the discipline to ensure that Bella wouldn't become one. Before I left, we took Bella to one puppy class together, and Charli continued the classes while I was gone. I knew I was going to miss Charli and Bella something fierce, and Charli promised to do all she could to make Iraq feel like it wasn't so far away.

When the moment came to say goodbye to both Charli and Bella, I felt a mixture of heartache and pride. I'm not sure if it was all the history lessons I'd had in school or the movies I'd seen that glorified military service, but as happy as I was to have this newly formed family, I felt truly proud to leave them in the safety of our home while I went off to join the fight to protect it. Through tears, I hugged Charli and held her close. I bent down and looked at Bella with sincerity and spoke with a fatherly tone: "Now, Bella, I'm going to need you to take care of Mommy for me, understood, baby girl?" I kissed her wet nose, and she tried to nibble mine. We all chuckled, and I stood back up, and Charli and I traded one last hug and kiss. "I'm coming back to you. I promise."

Charli sent me a letter nearly every day in Iraq, keeping me up to date on all things Bella. She sent pictures of Bella in every letter. I got so much mail that the lance corporal in charge of mail call started saying he had to "Go get Sergeant Kugler's mail." Charli even ordered personalized postage stamps with a photo of her and Bella on the back porch of our cozy Nebraska home. When Bella started losing her baby teeth, Charli sent me one as a keepsake, along with a few of Bella's tiny pink stitches from her spaying surgery. In turn, I mailed poems to Charli and dirty socks for

Bella to play with and remember my scent.

I treasured the letters I received from Charli. She and Bella were my family. I couldn't wait to get home to see them again. But first I had a mission to complete.

The first emotion I recall feeling overseas was pride. I felt it the moment my feet touched the sand in Kuwait, our first stop in what was technically a war zone. Safe in comparison to Iraq and Afghanistan, but a war zone nonetheless. I snapped a mental picture of my boots in the sand. From that point on, no matter how much or how little action I would see, my identity would be that of a Veteran of a Foreign War. We were prepped for this moment, and I took the responsibility very seriously.

From Kuwait, we flew into Iraq and unloaded our gear onto the tarmac of the Al Taqaddum airstrip. The scorching desert sun was so hot that the glue that held the treads of our tan suede boots to the soles started to melt.

We sorted out our gear and lugged it over to the areas where we were going to stay, tiny two-man trailer houses. We called them cans, but they were more like the Ritz compared to the group tents that my buddies who'd deployed years before me had

had to sleep in, let alone the holes in the ground that the grunts had to dig during the invasion. I shared the can with another sergeant who later became a good friend, Brandon Herman.

We began our first day of work, ironically enough, by locking up our rifles in a small Conex box, a steel shipping container. Since I was the most senior sergeant in my MOS, I was handed a notebook that read HEAVY EQUIPMENT FLOOR CHIEF across the cover, and we traded our rifles for training manuals and wrenches. Our enemies were broken pieces of equipment. I was twenty-five years old, leading a squad of twelve in the shop, and I had no idea what the hell I was doing. I wish I could say I stepped up and did one heck of a job, but I was in over my head. I was fumbling through paperwork that I barely understood, trying to make sense of it all, trying to keep my composure for the younger marines in my squad. I eventually butted heads with a staff sergeant who I felt was only making matters worse by playing stupid mind games with the marines. "Have your shop fill a hundred sandbags and build me a pyramid!" When I asked him why, he replied, "So I can make them tear it down, to demoralize them." I refused. This wasn't leadership, this was a power trip. As stupid as the games were, he was still my superior. He transferred me to the mine rollers section, where I finished my deployment. They needed a heavy equipment operator, and though I wasn't much good at fixing broken heavy equipment, I was pretty decent at operating it. So you could say it all worked out in the end.

Overall, I was proud of the job I was doing, but it was far from what I'd envisioned. We were in a war zone, supporting the ground troops, but living entirely different realities. The bullets, mortars, rocket-propelled grenades (RPGs), and improvised explosive devices (IEDs) were always outside the wire, always outside the concrete fence that separated our base from the rest of Iraq. I was a marine, but I felt more like a construction worker wearing camouflage.

It was something that happened outside the wire while I was in Iraq that changed the course of my story forever.

During our pre-deployment training in Lejeune, I came home twice to see Bella and Charli. The last time I came home, right before I shipped off to Iraq, my brother Mike drove me to my reserve unit in Omaha. What I'll never forget is how Mike put the car in park, turned his shoulders to me, and looked at me with his steely blue eyes.

"I love you, Bob," he said simply.

His face was hardened, his tone sincere. It was so clear how much he meant what he said. Mike was never one for showing his emotions. He was always a man on a mission. Emotions didn't serve much of a purpose. But sending his little brother off to a war zone may have opened that door. He'd already been twice to a war zone. He knew what it could do to a person. How it could change a man's outlook on life. He knew not everyone gets to come home.

"I love you too, Mike," I said in return.

On base, we had limited access to the internet, but I found a way to log on nearly every morning. I would skip morning chow, work out in the gym, and then get email on the office computer — endless pictures of Bella, which I loved. Now, by "gym" I mean tent full of old exercise equipment, and by "office" I mean wooden scaffolding inside an old Iraqi jet bunker. I would sometimes even be able to use the phone to dial a base in the States and be patched through to a civilian line. That call would almost always be to Charli. I'd tell her about the happenings on base, and she'd fill me in on the family and friend gossip back home. I'd ask her to hold the phone to Bella so that I could tell her that I loved her too and not to forget about Daddy. It was nice to have this little time to call home, as it made Iraq feel not so far away.

While talking to Charli on the morning of December 9, 2007, I was reading through my emails and received a message from my sister Amy, a message that sent a shiver through me. When I read it, I knew it was the message of dark waters. It read simply:

"Mike was in an accident. You need to call home."

7
A Bird to Chase

Bella and I have been on the road less than a month, and we are driving in the winter cold with the heater turned up. We reach Portland, Maine, park next to a bunch of old dock pilings, climb out, and gaze at the Atlantic, smelling the tang of salty air. It's only four P.M., yet dusk is already falling. The car lights are on and snow covers the ground. A bluish-gray haze drifts in the air. I watch Bella, the smile stretched across her face, and I'm so happy she is here with me. *This is it.* This is that moment I had imagined before we set out on our journey. The moment where I had envisioned what this trip would look like without Bella, if I had waited for her to pass before I set out on it, if I had ventured here alone. The thought had broken my heart and had become one of the driving forces that told me to take this trip now, while she was still alive. I wanted her with me through it all. And now, here we are. Rob, Bella, and Ruthie the Runner . . . exactly as I had envisioned it. I mean *this exact spot,* this exact scene. I've never been here before, never heard of the town, never seen pictures of the dock. Yet, somehow, I'd seen this exact moment in my mind, months before it happened.

We push north and into the solitude of Maine's Acadia National Park. No one is on the road for miles in any direction. We park at an empty trailhead and camp overnight in Ruthie, and in the morning snow lies thick on the ground and we make first tracks. Bella and I feel like the only creatures on the planet, until we spot deer in the distance, quiet and somber and gentle. Bella attempts a hopping chase, and the deer bound gracefully out of sight. When evening falls, we note it's the last day of December 2015, and Bella and I cuddle up in Ruthie, parked at a trailhead. We are far into the park, near the ocean shoreline, and our plan is to rise early in the morning to catch the first sunrise of the year. We fall asleep to the sound of a storm outside. Waves thunder on the shore, the wind moans in the night, and snow falls hard all around us. This is the first time since that phone call eight years ago that I've felt this content.

■ ■ ■ ■

In 2007, Mike had completed his service with the Marines. He'd been with 3rd Battalion, 5th Marines in Camp Pendleton from 1997 to 2001, then with the Marine Corps Security Forces, Kings Bay, Georgia, from 2001 to 2005, guarding nuclear submarines. He'd done his time and got out, and he'd joined up with Special Operations Consulting (SOC), an independent security company. He did two tours to Iraq with SOC as a civilian security specialist, cutting down the risks to Iraqi citizens and U.S. military personnel, and then he completed his contract with SOC. He came home to the States and looked to reenlist in the Marines again, but he was going to lose his "time in grade" in the infantry if he did, which would have been a hard blow because his five and a half years as a sergeant would count for nothing. He'd be starting over. So rather than go back to being a boot sergeant in the Marines, he took a job working security at the Air Force base in Nebraska. Just as he'd counseled me when I was signing my contract: "Infantry guys get out and work security jobs with no path for success." He quickly grew restless. The man was a warrior. Warriors don't do too well checking the ID cards of airmen for ten hours a day.

Then Mike found out I was finally deploying with my unit. He joined up again with SOC and headed back to Iraq for his third tour, to Ramadi, a town about forty miles away from Al Taqaddum. Just before he left he said to me, "Hey, I'm going to do all I can to see you over there. Can you imagine how awesome it will be, two small-town boys meeting up over in Iraq?" Not only was I finally getting the chance to deploy, but my big brother, my mentor, was going with me, only a stone's throw away.

The roads between Ramadi and Al Taqaddum were treacherous. I discovered we had supply missions that ran near Mike's base, so I tried to get on those so I could go see him. I even got my Humvee license to be a driver for the transportation company so I'd be able to go on convoys that would allow me to meet up with him. I was pumped not only to see Mike, but also to get out from behind the wire and see a bit of this country. Then it all got shut down. New orders. No one from our platoon could go out on convoys. It wasn't our job. We weren't drivers. We weren't riflemen. We were there to be mechanics.

Mike told me not to worry. He said he'd

try to see me at Thanksgiving. But that meeting didn't happen, and I was as disappointed as I was when he couldn't make it to my boot camp graduation so many years ago. He sent me a picture of him bottle-feeding a baby goat. I emailed back to him: "I'm naming this picture: Goat Fucker."

Those were the last words that passed between Mike and me, the last words I ever shared with my brother.

My alarm rings a half hour before sunrise. Bella peeks out from under her blanket. This time she seems to be the one asking for the snooze button. "That's okay, baby, you just rest while I get breakfast ready." I brew tea and make oatmeal for myself, which takes a few minutes. Then I pour Bella's dog food into her dish. Her ears perk up and she quickly emerges from the blanket and gives a long, deliberate stretch, along with a tongue-curling yawn. Though Bella eats dinner at five P.M. like clockwork, breakfast comes as soon as we wake up. I'm like her in that way; I just need to start the day off with a little food in my tummy. Bella devours her food, as usual, and I lift up her bowl and tilt it for her so that she doesn't have to chase the few bits of remaining kibble around the metal dish.

We climb outside, stretch, and hike along the coast in the dark to an area that seems in line with Cadillac Mountain. A large rock overlooks the ocean, and we sit and wait to take in the majesty of the new year. The sea is glass, calmed after the storm, and all around us the world is waking up. Sunlight begins to peer through the stratus clouds. Slowly, the sky becomes brighter. Snow covers the rocks on the beach, and tall evergreens form the skyline to either side. The sky turns pink, purple, and a golden glow spreads above and reflects on the ocean waters below as the spherical shape begins to crest the horizon.

And then the sun is up. Another promise of a new day. A new year. I hug Bella, amazed that we've made it this far, and ponder just how much of this new year she'll be able to see. We've far surpassed the original prognosis of her life expectancy, and she shows no signs of slowing down. Her breathing is still strong, her tail still wags fast and high, and her joy for life is evident.

I've brought my tripod to the beach to capture a photo of the sunrise. I use a timer to take a photo of Bella and me together. I've taken countless photos of her, but at the end of all this, I want a collection of the two of us, reminding me not just of how beautiful she was, but that our relationship, our bond, was something beautiful in itself. I click the shutter button and then walk onto the beach with Bella, talking to her and exploring the coast, allowing whatever the camera captures to be as natural as possible. When I return to check the digital screen, I see an image that is exactly what I wanted. The morning sky is afire with shades of pinks and purples. A man and his dog beside him, both looking into the ocean with a sense of wonderment, are silhouetted in the bottom left of the frame.

Yeah. That's us, I think. *That's who we are. Two adventurers, on a journey, in it together, until the end.* A sense of warmth cascades over me, knowing that this photograph has captured a moment in our relationship that will now last forever, long after Bella is gone. I pack up the tripod and camera, and we head to a coffee shop to offload the pictures from the past few days. When I open the image in full screen, seeing the deep connection and the layers of emotion, a feeling comes over me that I need to get out. I write this passage:

This morning, as I witnessed the sunrise over the Atlantic Ocean, I was struck with the true meaning of our journey. We are out here not just to survive, but to thrive.

I have battled with chronic depression and suicidal ideation for many years, and am here to tell others that there is hope. There are countless reasons to live and love this beautiful planet.

I've waited too long for someone else to give me authority to speak on this, yet I must have faith in my strength as a survivor. I aim to inspire others to continue to seek the light by sharing my own journey. I will not let a part of me be the definition of me.

I will continue to live, love, explore, and, by God, laugh without apology. I've got a lot of work to do, but acknowledging that is the key to understanding. And understanding is the key to freedom. Spread the Light like wildfire.

Writing this for the first time and sharing it publicly on social media has a physiological effect on my body. For years, it's been as if I've had a ball of stress in my lungs, restricting them. Once I have spoken this out loud, once I have shared this dark piece of myself, it's as if I've blown that ball right out of my lungs. Suddenly, I can feel them expand to their full capacity. Finally, I can breathe.

■ ■ ■ ■

Bella and I return to the ocean. I sit on a driftwood log, warm in the sunshine, feel my hand burrow into Bella's fur, knowing she offers me so much unconditional love. I snap the leash off her collar and let her go. She sniffs once. Twice. Then is off and running on her three legs. She plunges into the ocean, barking. Swims back ashore. Runs out of the water, shakes the water from her. She's so happy. She runs back over to me as if to say, "Daddy, let's do this again." And again and again.

I think: *I don't want to end this. I don't want to end this at all.*

This journey is saving my life.

Over time, I've learned there are simple steps I can take to keep myself safe. I keep pistols at home, but they are always locked, and the pistols are in a different place in the house than the ammo. I also don't drink with a loaded weapon nearby. People argue with me about this, but every service member I've ever known who's killed himself was drunk when they did it. We weren't allowed to drink in the field with live ammunition for a reason. It's dangerous. Nearly every time I've ever screwed up my life, I've been drunk. If you're battling depression and thoughts of suicide and you have booze and a loaded, easily accessible weapon at home, then you're setting yourself up for failure. It only takes that one moment of lapsed judgment. With a loaded gun, that's one moment you'll never get back.

Bella helps so much. When I've been stuck behind four walls every day — whether at home, school, or working a menial job — anxiety floods in, and it sets up its own movie theater inside the walls of my mind. But being outside with my dog where I can see everything around me, watching Bella take in countless new experiences with such verve as we travel together, seeing a humanity that isn't at war with others, this takes the anxiety away. This nature. This stillness. This silence. This incredible companion by my side.

This journey is saving my life. And the thought's slowly dawning on me, as time goes by, that it might just be saving hers, too.

Bella and I head south and drive through the Lieutenant William F. Callahan Tunnel on into Boston. We stop to pay a toll, and the guy behind the window says in this superstrong Boston accent: "Whelcome to

Bahston, dhood." Bella sticks her head out the back window and gives a little nod in his direction. We've stopped at so many tolls in the northeast that Bella expects treats now. The toll operator nods to Bella and adds, "Aw yeah, sweet dahwg, dhood." He tosses me a treat and I hand it to Bella, who woofs it down in a single gulp. We drive on into the tunnel, already feeling welcomed.

We hit the *Cheers* bar, run side by side through the Boston Common, and stop at Fenway Park, then head to New York City and trade open skies for billboards and honking geese for honking cars. At Central Park, Bella immediately sets off on a sprint. I run alongside as fast as I can to allow slack in her leash. "Don't let me hold you back, baby girl!" She picks up speed: "Try and keep up, old man!" We run and run, nearly the entire length of the park. Joggers, bikers, and walkers all call to us and wave, marveling at Bella's pace as her ears flop with each bounce of her front leg. Bella spreads happiness wherever she goes, and the returned smiles only fuel the happiness I'm feeling today.

Bella and I drive to Philadelphia. We reach the bottom of the famous *Rocky* steps at the Philadelphia Museum of Art. Bella takes one look at the steps and runs to the top. I can't help but play the *Rocky* theme music in my head. A group of kids stands at the top, and when they see Bella, they all let out an honest and completely innocent "Awwwww." I take in the scene. The children's faces aglow with love and admiration as this three-legged dog conquers these iconic steps like a true champion. I kneel, rub Bella's ears, and say, "Look at us, baby girl. We're doing our little bit to make the world a better place. You and me." This is our story now.

We stop for a quick peek at the Liberty Bell, and as we head back toward Ruthie, a man tries to sell us a horse carriage tour. "I'll even give you half off for the pooch," he jokes. I ask if he's serious about allowing Bella to ride, and he says of course. This adventure is all about *living,* so I load Bella up in the carriage, and as we ride along we hear some amazing history lessons from the driver, although Bella is more interested in the giant dog with hooves pulling us. I can only imagine what it looks like to passersby, but I honestly don't care, this ride is ours to enjoy.

From Philadelphia, we head to Washington, D.C., where I park in front of the National Museum of the American Indian and car camp overnight. The next morning,

Bella and I walk down the National Mall and see the Capitol building, the dome hidden under layers of scaffolding. As Bella leads me on her happy hop down the National Mall, we see an older African American man sitting on a bench. He says hello and asks if he can say hi to Bella. He's dressed in a tan overcoat, and his hair is slicked back and his mustache is gray. He takes nips from a bottle inside a paper bag and he tells me he's homeless. Bella sniffs him politely and allows him to pet her coat and scratch behind her ears. We talk a bit about the resilience of dogs and the unconditional love dogs can give. He takes another nip from his bottle, then launches into a speech. "That dog is all about love," he insists. His voice grows louder, and it adopts the powerful quaver of a Gospel preacher: "She don't judge me because I don't have a home. She don't judge me because I'm black!"

"Yeah, wouldn't it be great if we were all like that," I say, imagining a utopian world where we are all happy puppies.

"Now, wait a minute," he interjects. "You gotta remember, she's just a dog. A dog don't have no responsibilities. A dog don't hafta pay no rent."

His voice grows even louder toward the

end, and people start to stare. I'm interested to hear what he has to say, but I sense that our conversation has now changed to a monologue, and whatever's in his bottle has started to do the talking for him, because now he's on to some other topic entirely. He looks past me and begins to ramble to the masses. I offer a simple "Well, have a good day, sir." But he doesn't notice, so we move on.

Bella and I continue down the mall until we see the tall spire of the Washington Monument, where we sit on the grass to rest. I think about the man's words. "Remember, she's just a dog." I look into Bella's eyes as I pet her.

But you're so much more than a dog, I reason.

It's interesting, I think to myself as I gently stroke behind her ears, that at some point here, the tide has turned. I'm Bella's caretaker, I love her, and it's my job to keep her alive — that's always been the deal, since day one. But it might be fairer to say that for some time now she's been the one taking care of me.

Ruthie's brakes have been screaming since before Boston, and we're able to pull into a mechanic's shop for a diagnosis. She needs new brake pads and rotors, but the cost for labor is atrocious. I can change my own pads and rotors — I just need a place to do it that's not in the snow. Outside of D.C., I stop at a NAPA and buy the parts and post on Facebook: "Anybody in the D.C. area with a garage?"

One of my buddies from Team Rubicon soon responds, saying his brother owns a large tree-trimming business and has a shop where they do all their own maintenance on their vehicles. We reach the brother, who says, "Sure, bring it on over."

The brother is a military man himself and takes charge right away, putting two of his mechanics on Ruthie. They have all the right tools, and there's a couple of pins or collars I wouldn't have been able to get off by myself. The guys working on Ruthie are covered in grease and grime — the backbone of America — and as they work they ask me questions about Bella and our trip. They're finished in no time, and when they're done, I go to pay, but the brother says: "Nah, we got you covered. We're just happy to be part of this. You and Bella — what you're doing is inspiring. We want you to keep going."

I'm astounded. Grateful. Bella and I say thanks. I slip the mechanics forty bucks for beer money, and we adventure on.

Almost everyone we meet along the way lights up with energy. Everyone wants to see us continue. I don't know why for sure, but perhaps they crave setting out on their own journey. Their lives won't allow them to leave everything and *just go* like we have. So they offer to help us continue by purchasing photos or offering places for us to stay. Maybe it's so they can be a part of this story with us. Or maybe it's simpler than that. Maybe they see a dog in her final months getting to experience more than most dogs ever do. Maybe they see a man needing some time to figure out what his life is all about. Maybe they see that the two of them need each other and this journey, whatever this journey is, and they want to help. When we know each other's stories, when we see the love and the hurt, we relate to those things within ourselves, and the best parts of us come forward.

Our hearts.

On our way out of Washington, D.C., a heavy snow is falling, and the wind blows with ferocity. Forecasters are calling it "Snowmageddon." We're still in the city, and everywhere cars are sliding around, bumping into each other. Ruthie's tires aren't studded, but the treads are deep, and we put her in four-wheel drive and slowly push forward. We start up a hill, and Ruthie can traverse it no problem, but the cars ahead of me keep spinning out and sliding back down. Everyone is going nowhere fast, so we pull into a nearby parking lot and I roll down the window so Bella can keep an eye on me, and I run forward to help.

The first guy nods through his window, and I get behind the car and push with all I've got. My push is enough to help the car gain traction. He makes it over the hill. Behind him, a woman is stuck in her minivan. She smiles, but her eyes look worried, and I get behind the van and push. She makes it over the hill. I run back for another car. Another and another. Ten cars later, the road is clear. When I hop back into Ruthie, my pants are frozen. Sheets of ice cake my jacket, and Bella licks the frozen whiskers on my face with a wagging tail. I'm damn cold, but I feel good. Really good. One man and his three-legged dog can't save the world, but I can see to it that everywhere we go, we leave it a little better than we found it.

Bella and I drive to Fredericksburg, Virginia, and the snow is still steadily falling, and I fear that the roads might soon be closed, so we decide to drive through the night as Ruthie plows first tracks in the deep snow. In the rays of the sunrise, the sky clears, the sun comes out, and we reach the Shenandoah Valley. Above our heads, the sky is golden blue, and on the ground the snow lies bright and glistening.

In time we reach Camp Lejeune, North Carolina, a place that holds many memories for me. The snow's gone and it feels like a warmer season has begun. We meet up with my former commanding officer, Major Brandon Cooley, who lives at Lejeune with his wife, Gianni, their three children, and Rocky, their faithful ol' bulldog, gray in the face and slow in his steps.

Brandon and I started out together as reservists and were corporals at the same time. I had the honor of being one of the sword bearers at his wedding. Brandon pressed further into the military, and after he received his commission, he went on active duty, where he served as a combat engineer in Iraq. I was proud to salute him and call him sir. Brandon is a marine through and through. He looks like a G.I. Joe action figure and he's even fought a few pro bouts as an MMA fighter.

Now, so many years after Iraq, Bella and I

turn up at his house and I'm in awe of his continued service and dedication to the Marine Corps. Being a marine is who he is. Part of me misses being a marine, walking so tall wearing that uniform among brothers. There's really nothing like it. But was I ever really that? I feel almost as if I never really was. When I first came home, Brandon gave me sound advice during the "med board" process, where the military ended up medically retiring me. "Man, sometimes I feel like a shit bag for not fighting the medical retirement," I'd said back then.

"Kugs, don't think that for a second," Brandon said. "You did your part. You and your family have sacrificed enough."

I'd been battling a pain that had started near my sacroiliac joint on my side, ran around my hip, through my right testicle, and down my right leg. When my mind, body, and spirit were whole, it seemed I could tolerate this pain okay. But with a piece of my spirit missing and my mind in a million places, the pain goblins had ratcheted up the settings. The pain became intense. Unbearable. A constant throbbing. Each day I thought about pain more and more. Pain became the center of my existence. But no one was able to truly track the source of my pain.

Its origin a mystery, we scrounged for possible mechanisms of injury. One theory was that I'd taken a particularly flat-footed landing while jumping out of a military truck during training. I recalled it vividly. I'd felt a shock course through my back once my feet hit the ground, my body loaded with body armor and a full pack. Another theory was that I'd been lifting too much heavy gear when I should've being using a forklift. I'd bear down and manhandle equipment, trying to prove all my strength training in the gym had paid off. That's one of the many curses of testosterone — stupidity.

Today, we sit on Brandon's couch staring at a TV show neither of us is watching. The pain still exists, but I see Brandon in his uniform and can't help but think that if my mind were strong enough, I could be standing next to him in mine. I say, "I feel guilt, you know. Guilt for not fighting it."

"Nah, set that down," he says. "Don't focus on what you didn't do. Focus on what you did do. What you're doing now. What you have yet to do. You did enough."

Coming from my former commanding officer and lifelong friend, his words have weight. I keep hearing similar things from my peers and superiors, but why can't I accept this myself?

Bella and I stay for a few days, and I'm able to watch Brandon interact with his wife and children up close. I wonder if I'll ever be that family man, while Bella does her best to befriend Rocky, their elderly bulldog, who's not too keen on the new happily hopping guest. The family takes us right in, and we feel part of their family, which is welcome. Bella and I are a tight unit, but it's good to be part of something bigger sometimes. Their youngest son, Gabriel, was born an ultra-preemie and has a few complications with his right arm, so he's needed to go through a lot of physical therapy. The family calls his arm his "lucky fin," just like in *Finding Nemo.*

Gabriel is fascinated with Bella and her three limbs. There's an obvious connection between the two. Both of them have adapted to life with an impairment, but neither one of them are living impaired lives. Bella brings out happiness in most everyone she meets, but in these special interactions, it's more than happiness, it's relatability. It's encouragement. I take a picture of the whole family, and Gabriel sits right up front, hugging Bella. Bella loves him right back. When I go to leave, Gabriel climbs into the 4Runner along with Bella, cuddles her one last time, and says, "Wow, is this

Bella's house? Mom, come and look at this!" I smile as I once again witness the healing magic of this dog who has the power to untangle even the most complicated situations.

Bella and I head over to Onslow Beach, still on the base of Lejeune. The navy used to practice amphibious landings on the beach, but these days it's used as a recreation and camping area by folks at Lejeune. We climb out, and as I gaze into the ocean waves I'm transported back in time.

The first time I'd ever seen the Atlantic was from this very point, during pre-deployment training. At Lejeune, on the first free day we had, we all went to the beach for a barbecue. We played football on this same sand and swam in the warm ocean waters, and I'll never forget when one of the young marines found a shark tooth. We all scoured the beach for hours with our heads down looking for our own shark tooth to take home, but there was only one victor of the tooth hunt. Standing here today, I see the scene playing out in front of me, as if my memories are being projected onto the beach. *Wow, this really did happen. We really did go to Iraq.* It seems so surreal, as if these memories are from an alternate existence. I want to stay on this beach as long as we can. The emptiness of it seems a rare opportunity to have the beach to ourselves, and with it, ultimate freedom off the leash.

As Bella and I walk along the water, the more I feel connected with that former life. I keep my eyes peeled for a shark tooth, and I feel my buddies alongside us. Bella is hopping along off leash, sniffing whatever piques her interest, then suddenly her nose catches a scent that lifts her head into the air and leads her out into the ocean. Once she's chest deep, she stops. I quit looking for shark teeth and watch her with inquiring eyes. What does she smell out there? Nearly five minutes go by, and she remains in the same spot, looking into the water as the breeze blows over the waves and makes her ears flap in the wind. She looks back at me, then back at the ocean. I stare and wonder if she's hearing a call to go home — to swim into the ocean and never return. But she turns and comes running back to me. "No, I'm not ready yet, silly Daddy! We've got too much life left to live!" Today is not her day — and for that I am glad. We will continue to enjoy life tomorrow, but I wonder if she has any sense at all that her time is coming. And I wonder how lonely life is going to feel without her. I can't actually begin to imagine what it'll be like, and my spirits sag.

We stay one night in the 4Runner and the next day inquire about a strip of tiny cabins along the beach and find they run forty dollars a night. We book one night there and enjoy our own little slice of heaven. The next day we head back to Brandon and Gianni's house for one more evening. I tell them about our experience and how alive we felt out there on the beach, and the next morning Brandon winks at Bella, slips me a hundred-dollar bill, and says, "Why don't you guys spend a couple more nights in those cabins."

Bella and I end up staying in the cabins for four nights. During the daytime, nobody is on the beach except the two of us. Each morning, the alarm wakes us in time to watch the sun rise over the vast ocean. Oftentimes the sun rises from behind a wall of clouds that appear to be sitting on the horizon. Shades of orange and pastels fill the sky. We walk up and down the beach, combing for seashells and ever elusive shark teeth. Every so often we see other people, but rather than rushing to meet them, we allow them their space and keep within ours. It's a strange feeling, as I'm so quick to want to connect with other people. Yet here, on this beach with Bella, I have everything I could ever need. We take a few breaks inside, making ramen noodles and editing photos. Then we return to the beach to watch the sunset together. Every time we walk up and down the beach, I need to keep my eyes on Bella. If I don't, I see ghosts of all the young marines from my platoon running around, ready to head to war. I'm curious if today they struggle like I do, feeling as if they were never actually marines, that they were never really part of the fight. If they do, then I'd tell them the same words Brandon told me: "Don't focus on what you didn't do; focus on what you have done and have yet to do."

Bella and I have a lot we've yet to do.

I watch her hopping through the water without a care in the world as she gives chase to a low-flying pelican. Right now, there's a bird to chase. And right now, that's all that matters.

8
Flag-Draped Box

It didn't take long for Bella to learn how to catch and retrieve a Frisbee, and it became one of her favorite games. She learned how to leap into the air, open her jaws at just the right moment, and snag the flying disk midflight. Her tail would rise like a flag whenever she brought it back to me, and I'd reward her with lots of praise.

Over time, Bella got good at watching the trajectory of flight. I could even whip the disk at the garage so it would bounce off, and Bella learned to position herself to catch the disk off the bounce. Once I had a wild throw and the disk landed high on the garage's roof. Bella looked at me, and I looked at her. "Daddy. What have you done?!" she asked.

I got the ladder, placed it against the gutter, made sure it had a good footing, then started climbing. Bella stared at me with wide eyes. Once I got on the roof and started walking toward the disk, Bella started barking at me in a way I hadn't heard before.

A good dog senses danger. A good dog protects all those in her care.

"Hey! Hey! That is not safe, Daddy!" she seemed to be saying. "Get down from there, right this instant!"

As she barked, she ran back and forth, as if to alert the entire neighborhood of my perilous situation.

"Calm down, baby girl. Daddy's okay!" I called down.

She had no care for my words and continued barking, even putting her two front feet on the ladder. When I finally came down, disk in hand, she let out a mixture of ar-roos, barks, and little growls. I understood the implication. "I'm so glad you're safe, Daddy. But don't ever do that again."

I rubbed her ears and held her head close to mine. Bella's protectiveness astounded me. "Oh, baby girl," I said. "Here I thought I was the one looking after you."

That email from my sister read simply: "You need to call home."

In my mind flashed a thousand bad scenarios. Convoy drivers were having issues at night with depth perception and sometimes ran off the road. I imagined Mike in such an accident, perhaps somehow losing his leg. If the accident was already reported at home, then it must be serious. From Iraq, I called the base in Omaha and had them patch me through to Charli. My voice shook. Charli must have read the tone of my voice, because all she could say was "Have you talked to Melissa?" Mike's wife. "Just call her," Charli added.

That's all Charli said. I could tell Charli was uncomfortable, like whatever she might have wanted to tell me needed to come from Melissa, who is a former marine herself. I tried to keep an optimistic mind as I again dialed the base in Omaha and asked to be patched through to Melissa's cell phone. Melissa answered right away.

"Bob? Is that you?"

"Melissa, what's going on?"

"Mike . . ." Static. "Convoy . . ." Static.

"Melissa, I'm sorry, I can't hear you."

"Mike's convoy . . ." Static.

"I'm so sorry, Melissa, you're still breaking up."

"I'll . . ." Static. ". . . outside . . ." Static.

"Bob? Can you hear me?"

Finally. "Yes. I hear you perfectly."

This would become the fourth time Melissa needed to repeat this difficult information to me. Her voice sounded heartbroken yet calm. She said simply: "Mike's convoy hit an IED. He didn't make it."

"Wait . . . what?"

"He didn't make it, Bob."

"Mike's gone?"

"I'm so sorry, Bob. So sorry. He loved you so much."

I have no recollection of how the conversation ended. I stood from the chair and made it three steps before my back found a wall. I slid down until my butt hit the floor.

Details were sketchy. A roadside bomb had exploded, taking out the vehicle Mike was riding in. They were southeast of Baghdad, down near Al Kut. Two other security specialists were killed in the same blast — Texan Steven Evrard and Washingtonian Micah Shaw, a father of three young children. Another man, Billy Johnson, was seriously injured. Mike was in an accident? Their deaths were no *accident.* They were killed by an improvised explosive device planted in the road with the intent to kill.

Corporal Welch, a towering man who we always gave shit to for being part oak tree, saw me crumpled on the floor and ran over.

"Mike . . . my brother." I could barely get out the words. "He's . . . he's *dead.*"

I needed to get home. My family needed

me. Still in a blur, I stood up. I knew I would need a Red Cross message to begin the necessary paperwork to come home. I ran and found Gunny Ortega, our staff non-commissioned officer in charge, who took me to headquarters, a one-level office building that sat between two jet bunkers. Our command staff, Major Moen and Gunny Peacock, tracked down the Red Cross and started the orders for my emergency leave. In all the commotion, Gunny Peacock received a message that felt like a hit across the chest.

"Mike's here," he said. "His body is on our base."

Turned out, Al Taqaddum was home to the morgue for all the allied forces fighting in the Al-Anbar Province. Mike's body was on my base, along with the two other men.

I said simply: "I want to escort him home."

The command staff agreed and left to make arrangements. A chaplain was in the room, and we spoke for a while. Then the command staff returned with another message: "I'm sorry. His body is already gone. He's in Kuwait by now."

Mike had been so close, yet we'd just missed each other. I knew I had to catch up with him. I needed to be there with him. I needed to be the one to bring him home.

I ran to pack my gear. My good friends Lance Corporal Steward and Sergeant Weddle stuck by my side, partly because they wanted to, and partly because Gunny Ortega had ordered a 24-7 battle buddy for me — a guy who'd watch my back and care for me during crisis.

An extra seat was available on the next flight out. I rushed over to the airstrip. A few tents and simple structures provided shelter from the desert sun and whipping sand. I made it just in time for my flight, but then received devastating news. An officer had ordered the gear on the plane to be rearranged. My seat was gone. I'd need to wait. An hour passed, and then another, and then another. Night fell, and I continued waiting in one of the tents near the airstrip. The sun rose. Morning passed, and afternoon began. Still I waited.

A first sergeant walked by. He recognized me from training in Quantico. "KUGS!" he said, with the excited tone of seeing an old colleague. "What's up, man?"

I told him about my brother and how I was hoping to escort him home. The first sergeant's tone changed to all business. He had a friend who worked the morgue and promised he'd track down some information for me. Fifteen minutes later he returned. "Kugs! I talked to my guy. They just left about three hours ago. They're on their way to Kuwait."

"They *just* left?" The earlier information I'd received more than a day ago had been a mix-up. Discouraged, even angry, I asked, "You think I can catch him in Kuwait?"

"Well, man, it's not looking good. They're due to fly to Dover, Delaware, tonight, and we can't hold all three bodies back. We can't hold just him either, because we don't know who is who yet. They won't be identified until they get into the States."

I thanked him. What else could I do? I went and phoned my mother, walking her through this news. The information about Dover was particularly tricky to convey.

"Why don't they know who is who?" Mom wanted to know.

How was I supposed to explain this to our mother without being horrifically graphic? "They just . . . don't know, Mom."

I sat down by the airstrip and waited some more. Minutes felt like days.

Late that evening, a spot on a plane finally opened. I buckled up on the red cargo netting seats of a C-130, and a couple of hours later we landed in Kuwait. I exited the plane onto the tarmac, and for the first time in my military career, I felt completely alone.

All right, Kugs, you got this. Turn in your gear, then find Mike. I wandered alone through Ali Al Salem air base, looking for the gear depot. I knew I had to turn in my Kevlar helmet, flak jacket, and M16 rifle. I found the right area and removed the three-point sling from my rifle. The sling was made of olive drab webbing and had KUGLER stenciled along the outer strap. Mike had sent me this sling; no way was I going to give it up. It felt eerie, not having my rifle. I had cursed the damn thing for months, lugging it around with never a need to use it, but now I felt naked without it. I asked around and was pointed in the general direction of the morgue. Walking alone on base with no rifle felt surreal, but I could move at my own pace, a brisk pace, the way Mike used to always move, and I could never keep up. No time to waste.

I made it to the morgue. Mike was already gone. *Damn it.* The news felt devastating all over again. My only chance now was to catch up with him in the States. I called the Charles C. Carson Center for Mortuary Affairs in Delaware, where the remains of service members and government officials killed overseas are sent to be processed, identified if needed, and returned home to their next of kin. I was told to head home and wait until dental records were received. They'd call me back in the States. Even then, it would take a few days. I'd done all I could do.

I headed toward what would become my next flight — a civilian jumbo jet flying out of the army base — and prepared to wait some more. Finally, I boarded the plane for what felt like the longest ride of my life. We flew across the Atlantic and landed in America. I'd been up for more than two days and two nights. Everything felt blurry. A couple of connections later, I finally arrived at the small airport in Lincoln.

Charli was there waiting for me, and she wasn't alone.

Avery was there. Nick Peterson. Mueller, Saccomon, Yashirin, and more — my entire crew, all marines from my unit.

I've never been wealthy, but I've always been rich in friends. And when the shit hits the fan, you go to your friend's side and you stay there. That's all it takes. Just being there. That's exactly what my friends did. They showed up.

As soon as Charli and I got back to our house, Bella bolted out of her kennel toward me. I was amazed at how much she'd grown. "Remember me, baby girl?" I asked.

She ran in circles, shook her rump, and licked my face with such happiness: "Of course, of course I remember you. You're Daddy!" Her obliviousness to what was going on was beautiful yet poignant at the same time. I wanted to snuggle and cry with her; she wanted to play. Still a puppy, her mood-sensing skill hadn't quite kicked in yet. Or perhaps she was the only one that really knew what I needed, which was a good game of tug-of-war with a dirty sock. "Okay, girl, let's play," I said. Before I could grab the sock, Bella scooped it up and ran around the coffee table, through the kitchen, in and out of the bedrooms. I reached for the sock and she let out a playful growl: "I dare you to come and get it!" Finally I dove toward her and grasped the sock in my hands. Tug-of-war quickly turned into a wrestling match. We were battling over this dirty garment as if it were the last piece of sustenance on earth. The distraction was exactly what I needed. Thoughts left my head and laughter filled the room.

We were three rounds into the game when suddenly I remembered something. I picked up Bella off me, ran to the dresser in the back room, and located a sealed manila envelope that read: "To be opened in case of funeral." Mike had left this envelope with me before his first deployment years ago. Though I had hated hanging on to it — hoping never to have the need to open it — I was so thankful to have it now. Grasping it in my hands had a ghostly feeling, knowing that the words on the outside were written by his hand, and that by breaking the seal, I'm opening his last wishes. Inside, I expected to find a will, instructions, a detailed list of what to do, the kind of thorough arrangements and organization that Mike excelled at and enjoyed. But Mike surprised me, he surprised us all. Inside was a heartfelt letter, with segments dedicated to each person in the family.

To Melissa, he spoke about how he couldn't bear the thought of leaving his beautiful bride too soon, of not being there for all the wonderful memories they would have shared in the future. He described her as the love of his life.

To our mom, he thanked her, calling her a hero and a role model.

To our brother John and sister Amy, he commended them for doing well in adult life, for being great parents to their respective children. He was proud of them.

And then he wrote,

To my little brother Bob,

Oh, how this letter gets harder and harder to write. I miss you already, man. Don't tell anyone, but I cried like a baby the night you graduated boot camp and I couldn't be there, because I was stuck in the middle of [fighting fires in] Idaho. I was so proud of you.

You've got more talent in your little finger than I do my entire body. Follow your dreams and go for it.

Love you, bud. Take care of things for me.

For the first time since hearing the news, I dropped my armor. I let it all out. Charli held me in our bed as I cried. The pain came out in the form of a wail.

"I'm going to miss him so much," I blurted, the words barely recognizable.

The world wasn't ending, but a large piece of mine just had.

What the hell was life going to be like now?

I had only been home a day, but we lived nearly two hundred miles from my mom, and I needed to get home to her. I said goodbye to Charli and Bella. "I've got to go do something important," I whispered to Bella. "I'll be back much sooner this time, I promise." I climbed into the car alone and drove to my mother's house in Broken Bow. The pain of losing my brother was indescribable, but I can't imagine what it was like for mom to lose her son. Mom met me at the door, and we hugged for a long time. We sat down on the couch and talked about grief, about how grief can come in waves, sometimes far after the event that initially caused the grief. She hadn't cried yet. "Why can't I cry?" she asked. "Don't worry," I said gently. "You will."

I got in touch with Mike's best friend from his time in the Marines, Josh Schulz, and I asked him to come with me to Dover. He said he'd already packed his bags. We flew to the East Coast the next evening. But something was nagging me.

There won't be a flag.

Mike had been a marine for eight years and served three deployments overseas, where he provided security for the Army Corps of Engineers. But since he was a contractor at the time of his death, technically he was killed as a civilian. The mortuary officials informed me that since Mike was no longer a service member, Mike wouldn't be covered by a flag. This made me livid. Mike's service *mattered* to our country. It mattered when he was a marine, and it mattered when he was supporting our military as a contractor. His death mattered, because his life mattered! When Josh and I arrived in Dover that night, I asked an airport official about the policy. "Honey," she said. "When you get down on the tarmac tomorrow morning, you can do whatever you want." I breathed a big sigh of relief. Until I realized . . . I hadn't thought to bring a flag of my own.

Josh and I took a cab to our hotel. It was ten P.M. by then, and on the way over, I called every store in the area. Nothing open. I kicked myself for not thinking of this earlier. We set our alarms and went to sleep, but I awoke two hours later. *A police station!* Police stations have flags. I found the number and called. No flag. The officer recommended the United Service Organizations. *Of course!* It was nearly two A.M., and some USOs run around the clock, so I found the number and dialed. They had a flag — and yes, we could have it!

"I'll be there first thing in the morning!" I said. Our flight back to Nebraska left at seven A.M. "We'll make sure to get there in time."

My eyes had barely closed before the alarm rang. Josh and I donned our dress blues uniforms, making sure all the eagles' wings on the buttons lined up parallel to each other. We arrived at the airport and made our way to the USO, located inside. Sure enough, there was the American flag. But it was so high on the wall near the top of a vaulted ceiling that it couldn't be reached with an ordinary ladder. "We already called maintenance to bring in a scissor lift," the man from the USO told us.

We waited. No scissor lift came. Finally, we had to leave for our flight. The USO volunteer took my cell number, and Josh and I headed through security. Josh and I were asked to take off our dress blues jackets, because there's so much metal on them, and were ushered to a private room to do so. My phone rang. The USO guy said, "I'm so sorry, we won't be able to get the flag down. But I've called my boss, and she says there's a new flag around here somewhere. I can't find it. Today's her day off, but she's on her way to come get it! I'll call you again when she gets here."

It was six in the morning. I was so touched that this woman had given up her day off and driven to work at the crack of dawn so she could find this flag for me. For Mike. Josh and I got through security and waited. Our plane started boarding. Still no flag. The plane was halfway boarded by the time my phone rang again.

"We have the flag, can you come get it?"

Can we come and get it?

He said something about not being able to get through security. We didn't have time to get there and back before the plane boarded. I turned to the nearest TSA agent and asked for help.

"No," he said. Flat denial.

"No?" I replied, with a quizzical look on my face.

"We're not allowed to delve into third-party matters."

My jaw clenched as I turned around and looked at Josh. *What are we going to do?* Josh said he would run and get the flag. My job was to try to hold the plane, which had now fully boarded. Shit. Josh sprinted in the direction of the USO.

"Sir, we're going to have to shut the gate soon," the woman taking the tickets said.

"My bud will be back shortly," I said.

She tapped her foot. "I'm sorry, sir, we can't hold the entire plane for . . ."

I smiled and in a polite, low, calm voice explained the entire story to her. I took my time.

She listened and tapped her foot again, but I wasn't budging.

"Sir," she said. "You really must board the

plane — now!"

I hesitated. Just then, far off to my left, I heard the beep of one of those airport carts. It flew toward us, with Josh in the passenger seat. The cart's yellow light flashed as people got out of the way. In Josh's lap sat Old Glory. She was folded into her famous blue-and-white triangle. The blue-and-white triangle that signifies service. Sacrifice. Sorrow for those who remain.

As Josh held the flag, tears streamed down his face.

My eyes welled up at the sight, and I turned my head to look away, not allowing a tear to fall. *Stay strong. Stay strong for Mike. Take care of things for him.*

The airline officials took us down to the tarmac and directed us toward a lone luggage cart. The contents of the cart were shielded from our view by a dirty plastic curtain. Josh and I walked toward the cart. I took a deep breath and pulled back the curtain. I wasn't prepared for what I saw next. It was not a coffin. Nor was it the familiar metal box you see in photos. It was cardboard. A cardboard box with a slip of wooden frame around the bottom. On one end was printed a single word.

HEAD.

He's labeled like generic packaging. No. You don't do this. I wanted to find whoever was responsible, grab him by the scruff of his neck, and rub his face in the mess he'd made. But what good would that do? Pragmatism took over and I asked, as if to no one, "The stars go on the head?" I'd done countless burial details during my time as a marine reservist and had had the honor of folding the flag at a few. The direction of the flag seemed the question to ask. It was the only thing that made sense.

"Yeah," Josh replied. "I'd say so." His voice sounded far away too.

We draped the flag over the box, folded the excess fabric to make it presentable, and picked up the flag-draped box. It felt heavy enough to indicate we had most of Mike with us. We set the box on the conveyor belt that fed luggage into the belly of the plane. As the belt slowly started to move Mike up the ramp, the flag came untucked, and we rushed to re-tuck it. I cringed to think the flag was being pressed against the dirty rubber of the belt. *But at least we have a flag,* I thought. *Thank God, we have that flag.*

When we boarded the plane, the passengers came to a hush as two marines in their dress blues filed down the aisle. The pilot asked me the name of the deceased, and I told him. A steward ushered us to two open seats in first class — a kindness we hadn't expected. The pilot came on the loudspeaker and announced we were carrying a service member below, that his name was Mike Doheny, and that we were escorting him home.

We flew to our next airport, got off the plane, and headed down to the tarmac. We wouldn't be leaving him for a second. Mike came off the plane, and we escorted him to the holding area, waited with him there, then got ready to load him onto the next plane.

This time, instead of a well-worn luggage cart, the airline representatives directed us to a chariot. At one time it had been a luggage cart, but now the cart was freshly painted in a deep navy blue and bore the insignia of each branch of the military. Crisp, clean curtains shielded the contents with dignity. We placed Mike in the cart and ushered him across the tarmac into a larger hangar. One of the workers explained that a man had lost his son in combat, then had the experience of escorting his son's body on an old luggage cart. The father had commissioned the new cart, so that nobody at this airport would ever again have to use an old one.

The plane arrived. We escorted Mike to the plane and again placed him on the conveyor belt. Then we flew to Omaha.

When the plane landed this time, the pilot asked everyone to remain seated until Mike's body was removed. Josh and I exited the plane directly onto the tarmac. The cold Nebraska air sliced through my dress blues jacket. I walked to greet the marines from my unit who were there as the official honor guard. I shook their hands and thanked them. In the windows of the terminal, I spotted Melissa and her mother. They started heading toward us, and a representative let them through the doorway. Both Melissa and her mother appeared stoic, each staying strong for the other.

Ceremonial lines of marines formed on either side of the conveyor belt. Mike's flag-covered remains entered our view. Melissa's rocklike foundation seemed to falter. She knew Mike was gone, but seeing her husband's remains emerge from the belly of a plane, covered in the American flag, that made it real.

Mike and I had gone to Iraq together, and Mike was returning home in a box. I could not imagine the hurt in Melissa's heart. And what if she had seen that cardboard box with that one word written on the end?

Thank god we had that flag.

I hugged Melissa. Through tears of my own, I said I was sorry. So very sorry that I'd come home on my own goodwill, and that I was bringing her husband home in a flag-draped box.

Melissa, Josh, and I loaded up into her RAV4, got behind the hearse, and started the nearly four-hour-long drive back to Broken Bow. As we reached the outskirts of Omaha, I received a call from an unknown number. The caller identified his name and said he was from the Patriot Guard.

"It's too cold for our bikes today, but we're right behind you in a Suburban," he said. "If you need us, we'll be there."

The Patriot Guard Riders are a community of mostly motorcyclists formed to support military families during funerals. I'm a member myself and have rode among a hundred motorcycles escorting a fallen service member home. It was good to know the Suburban was covering our six.

Once we reached Broken Bow, we helped unload the sacred cargo, the flag-draped box, at the funeral home. Then we could do nothing more but head home and get some rest for the night. Josh and I returned to the mortuary the next morning. Mike was now placed in a beautiful natural-wood casket that was covered with a new, clean flag. The flag we'd initially draped over Mike was folded neatly, although now dirty from contact with conveyor belts. I asked Melissa which of the two flags she would like, and she chose the one we'd covered him in, since it held more sentiment to her. We took the flag to the dry cleaner and asked them to clean it meticulously and handle it with extreme care. Then we brought it back to the mortuary and replaced the new one with it, since during the funeral service Melissa would be given whichever flag was on the casket. Our mother would be given the other one.

As we exchanged the flags, I asked Josh to help me place the cleaned flag directly over the top of the other one. I held the top flag by the stars as we pulled the second flag out by the stripes at Mike's feet. I did not want Mike's casket to be without the flag for even a moment. The sight of the two flags made me swell with emotion, and I hugged the red, white, and blue coffin and sobbed. "I love you, brother," cracked through my vocal cords as Josh placed his hand on my shoulder: "He loved you too, Bob."

The day of the service was bitter cold, but

it did not stop dozens of members of the Patriot Guard from showing up and forming a flag line in front of the building. Guard members stood side by side as they each held on to a pole with Old Glory flapping in the icy wind. My mom walked up and down the line, thanking them and shaking the hand of each person. That's my mom. Smiling and thanking everyone else, even at her own son's funeral.

The service was filled beyond capacity. The entire community had come together to honor Mike's sacrifice. I whispered, "Look at this, Mike . . . they're here for you, brother." As I looked at the packed crowd of somber faces, including nearly every teacher from our school, members of the police and fire departments, friends and family, I smiled knowing that they had long forgotten or forgiven any teenage misdeeds and came today to say goodbye to a man they respected.

From the front of the auditorium, I read Mike's letter. The letter showed Mike's heart for his family and friends, for love and service. He likely had no idea the size of the audience that his letter would be read to. Funerals in our family rarely filled a room, yet today I stood before an auditorium filled to capacity. My knees began to quiver as I surveyed the crying faces in the crowd, but I got through the letter, I shared Mike's last words with our entire town. Kevin Cooksley, the rancher who was Mike's mentor, also spoke, as did both of Kevin's daughters. The loss of Mike was felt in their words. I'll never forget Kevin's voice wavering as he said: "Now I know why Mike always walked so fast. He had so many places to go, and so little time to get there."

As if that weren't enough to send shivers down my body, the bagpiper began to play "Amazing Grace," a soulful sound that can pierce through the strongest emotional armor.

Josh and I and a handful of my marine buds were the pallbearers. We carried Mike out to the hearse. When we pulled into the local cemetery, marines from my unit were standing motionless at parade rest, wearing thin cotton gloves, gripping the cold steel of ceremonial rifle barrels.

Final words of remembrance were said.

Balloons were released by my niece and nephews.

The twenty-one-gun salute blasted through the air.

The flag was removed from Mike's coffin and folded. The first sergeant walked over to Melissa and recited: "On behalf of the

President of the United States, the United States Marine Corps, and a grateful nation, please accept this flag as a symbol of our appreciation for your loved one's honorable and faithful service."

And then it was over.

It was time for me to return to Iraq. I could write a request to stay home, but I wanted to finish my deployment. I was sure Mike would've wanted the same.

I said goodbye to family in Broken Bow. It was particularly difficult to leave my young niece and nephews.

"Don't worry," I said. "My base is far from danger. I promise, I'll be okay."

"They've heard that before," Mom said with worry in her voice.

I gave my mom a big hug, looked into her eyes, and said: "Mom, I'm coming back soon."

She hugged me again. "You better."

I felt selfish for leaving my family, but they understood. They knew how much this meant to me. They knew this was what Mike would've wanted.

Back in Lincoln it was time to say goodbye once again. I took Bella outside and watched her romp in the snow without a care in the world. I called her over to me and bent down to rub her ears. I put my forehead next to hers and soaked up as much of her contagious joy as I could.

The contact with Bella was therapeutic. When you're stressed, depressed, anxious, or lonely there's a healing power to your dog nudging her nose into yours or licking your face. Already, Bella was proving to be an amazing therapist — she couldn't get enough ear massaging, belly rubbing, or booty scratching. And what she saw now was my need to be brought out of the moment I was trapped in and back into the moment in front of me. I petted her and looked into her eyes and saw this absolutely perfect creature in front of me. "Don't worry, Daddy, I love you, that's what matters," she seemed to say. I was grateful for the reminder. I needed to believe that it was all that mattered.

"Daddy's got to go again," I said eventually. "Take care of Mommy for me, baby girl."

"Just promise me you'll be safe," Charli said.

I straightened up, hugged and kissed Charli one last time, then headed for the airport.

One thought kept nagging at me, and I didn't catch its full significance until the adrenaline of the trip began to wear off. I found myself exhausted at this thought — not only physically, but mentally and emotionally. This thought was too much to bear.

It had been planted in my mind before the funeral, when we'd had lunch with Mike's coworkers from the security company. They'd told me more about what Mike was doing in Iraq. Their specific task was to provide security for the Army Corps of Engineers while they cleared and disposed of munitions. They played a direct role in the war on terror. At the lunch, a good friend of Mike's had informed me that Mike had taken a mission he was not supposed to be on. It was not his rotation, yet the route was to swing through Al Taqaddum, where I was stationed.

The friend looked at me across the lunch table and said, "Mike was doing all he could to make it to see you, Bob. That's why he took that mission."

Those words hit hard. It wasn't blame; I knew he meant it as a compliment — that Mike loved me to the end. And, sure, I saw Mike's devotion to his little brother by taking that mission. But I also saw the events lining up in an undeniable order. I wasn't playing the *what if* game. I wasn't casting blame on myself. Yet there was an incontestable truth I couldn't get past.

Mike took a mission he wasn't supposed to take.

Mike was killed on that mission.

Mike died . . . on his way to see me.

9
I Will Never Leave You

Bella and I are on the road, and as she lies on the seat beside me, I wrestle with all these thoughts. God, it's going to hurt so much to lose her, too. I reach down, stroke her ears, feel her coat. I'm reminded that she's here, right now. I shake the pain of her future death and get back to the joy of her current life. Already I begin to feel better, to breathe easier.

We stop in Savannah and end up staying an entire month in a historic Victorian home owned by my buddy Josh. He's in the military and away a lot, and Bella and I do odd jobs around the house for rent, since Josh refuses cash. Every day Bella and I walk to nearby Forsyth Park, where she spreads smiles to everyone who sees her happy face as she hops alongside me. More than anything, Bella just wants to be by my side. She listens to me not out of fear or a sense of duty, but because of a bond that is difficult to put into words and that I find only other people who have had a strong relationship with their dog can grasp. She's my dog, and I'm her human. We're a team. It really doesn't matter what we do, as long as we do it together.

Forsyth Park is full of eclectic people. A young African American woman with purple glasses knits a bathing suit. A group of young men play hacky sack. Bella and I approach a group of photographers and discover that most are from the Savannah College of Art and Design. I start talking about photography with them, and Bella takes the opportunity to say hello to another group of students nearby. And by hello, I mean she lopes straight through a circle that's gathered around a live subject for an art class.

I run over and apologize. A young woman's pencil sketch is now wrinkled with Bella's dirty paw print. The girl laughs it off, and Bella is oblivious to any wrongdoing as her tail wags strong and proud. The entire group breaks their concentration to gather around this amazing three-legged dog. The students are all younger than me by at least ten years, and they look happy to be out in the park, happy to meet Bella. But when I step back for a few moments and watch the circle crowded around Bella, I sense the weight of the world in their lives already. Lines of worry etch their faces. Creases of concern mark the corners of their mouths. The world is hard on everybody, no matter what your age. All the students keep saying hello to Bella. They pet her coat and scratch behind her ears, and Bella says hello right back. I step back into the circle. Once I tell them what we're doing, where we're going, the students love on Bella even more.

"Wow, just think of what you guys are doing," a guy says. "Just you and your dog and the open road. That's the dream."

Before Bella and I leave, I'm sure I notice that the students' lines are smoother, their creases less intense. I can't help but think: *Everybody needs a dog like Bella.*

February turns into March, and I learn that a weeklong Project Hero event is scheduled for the following week. Project Hero is a nonprofit organization that helps veterans rehabilitate their injuries, physical and/or mental, through cycling. This event, called the Gulf Coast Challenge, is a bicycle ride from Atlanta to New Orleans. Atlanta is only a three-and-a-half-hour drive from Savannah by car, so I'm simply too close to pass up this opportunity to see good friends who were once a huge part of my life.

My plan is not to ride the entire event, but just to pop up there and ride the first day, from Atlanta to Fort Benning. My good buddy Felipe, a former army drill sergeant, lives nearby and says Bella can stay with his family while we ride. I hate the thought of leaving Bella for an entire day, and I've vowed I will never kennel her again. But this is not a kennel, and she's in great hands. Any time I've ever been around people from Project Hero, I always change for the better. Cycling long distance is a form of meditation for me, and finishing each day alongside old friends and fellow warriors fulfills a need for purpose.

Sure enough, when I walk into the hotel in Atlanta, I'm instantly welcomed with hugs and smiles. I meet up with the founder of Project Hero, and he says: "Great to see you, Rob. Are you on for the whole ride?"

"No, just today," I say.

He looks at me closely. "No. We need you here, Rob. You need to be here, for the whole ride."

The entire challenge. Oh boy. I make a call to check that Felipe's family is okay to have Bella for an extra four days, and they're good. Everyone should have a Bella, and now they do. The real problem is that I haven't ridden anything longer than a twenty-mile bike path in two years. We're scheduled to ride nearly fifty on the first day of the challenge — and each day after that is more, totaling nearly five hundred miles in six days. I've put on a few extra pounds and the nerve pain in my leg throbs every day. I remind myself, however, to be thankful even for having these body parts that hurt. Many of the veterans with Project Hero are missing limbs, paralyzed, or even blind. Who am I to complain about a fully functioning limb that merely hurts?

The next morning I'm up at six, surrounded by a crew of veterans, all riding toward New Orleans. We head out at a brisk pace. There's not a lot of climbing at first, and it's not a race. Everybody's in this to finish together. I focus on each pedal stroke, each surge to the top of the hill, and ride in the middle of the pack, hoping I can push through pain and finish. Several riders around me are paralyzed or missing legs, and they ride in specially adapted cycles, using their hands and arms. I'm astonished by their willpower. Their motivation fuels my own. Every so often, a call comes for a "pusher" — someone who helps another rider over a hill. When a call comes, a pusher pulls out of line and rides forward. He grabs a special push bar on the back of a hand bike and uses his power to help the other rider up and over, then falls back in line again. I'm not in good enough shape to push. Yet when the calls come, I can't hang back. I push every chance I get.

As I ride I think. *Pedal . . . shift . . . dig . . . reach . . . boost the rider ahead of me . . . draft so I can rest. Feet . . . legs . . . knees . . . back . . . shoulders . . . everything hurts. I want today to be over. I want to stop, and we're only twenty miles in.*

A call comes, and I pull out of line. My hand finds the push bar. The navy veteran cranks vigorously with his massive arms. His legs lie motionless, small and atrophied. I drive my legs down harder with each stroke. The pain in my right leg is nothing. Up the hill. Back in line. Recover. Another call comes . . . again I ride forward. Again, I push with all my might . . . get him up and over . . . then fall back. Pedal . . . shift . . . dig . . . here comes the bridge . . . another call comes . . . I catch up to another buddy, help him push other riders up the bridge. Fall back.

Day two passes. Day three. Day four.

On day five we set out, riding toward Gulfport, Mississippi. Once we get there, we'll have about 350 miles behind us and one final day of nearly 100 miles to go. I'm ten miles in. Twenty. Thirty. I try to dig deeper but find I have nothing left anymore. I push in short bursts and can barely keep up. I can't push the guys who need help any longer . . . I can barely push myself . . . my legs ache . . . my back is on fire . . . I don't have the strength. We come to a massive overpass. It may only be half a mile, but it looks like it goes on forever. "Pusher!" I grit my teeth and push myself forward out of the pack and reach one of the hand cycles. My palm finds the push bar, but my legs have nothing left. I fall back. I ride so slowly it feels like I'm rolling backward.

I have failed.

I ride slowly . . . so slowly . . . my mind jumps to the darkness. Self-doubt wreaks havoc on my thoughts. I try to shake the negative thoughts out of my mind, but they are so loud: *You shouldn't even be here right now. . . . Why are you even trying? . . . Give up. . . . Quit. . . .* I can't silence this voice. I feel he's right. I should quit. I search through my mind like a catalog, looking for something, anything to get me back in the game.

An image of Mike's face appears. *I love you, bro. . . . You got this.*

Brandon, my commanding officer, flashes in my mind. *Don't focus on what you haven't done, focus on what you have done.*

Finally, an image of Bella hopping happily up mountain trails on three legs with a smile rather than a complaint. The decision to take Bella's leg wasn't an easy one, but seeing how she responded to the surgery assured me that it was the very best decision I could have made for her. "What kind of life will she have on three legs?" I'd repeatedly been asked. Well, she stood up on her own right after surgery. She was wobbly, but she pushed herself to her feet and hopped her first three-legged hop down the vet's hallway toward me. The vet was astonished. Day one after surgery, and she was already running around the house. Now she's climbed mountains, run across Central Park and up the *Rocky* stairs in Philadelphia. It's this image that propels me over the top. She's waiting for me at Felipe's house. Waiting. For me.

"I've got this," I whisper to myself.

I push forward. I catch up. I finish with the pack.

That night, thunder and lightning blankets Gulfport. Rain falls in torrents. The next morning, fish are literally on the road. Rain won't stop us, but lightning will. The last day's ride is canceled.

When I return to Felipe's place, his kids

let Bella outside and she runs at me, barking, slurping, kissing. She jumps at me and nearly knocks me over, then tears circles around the yard, barking, barking. I kneel, and she runs and tackles me again. I fall to the ground, and she pins me down, her lone front leg pressing on my chest as she licks my face with 50 percent love and 50 percent scolding. "Oh, Daddy, I missed you so much! Don't ever leave me again!"

I laugh, wrestle her off me, sit up, and say, "Oh, baby girl. I missed you too. Don't you know I was always in your heart?"

She settles down and leans into me as I rub her ears. She calms down and curls up in a ball on top of my legs. "And I'm always in yours."

On the way back to Savannah, Korrine, the sister of one of my best marine buddies, sees that Bella and I are in the area and invites us to spend the day with her and her husband and kids in Phenix City, Alabama. It's sweltering, and we all ride together in their car to the river to cool off. Instead of sand, the river's flanked by large smooth slabs of rock with tide pools to lounge in and natural slides to slip down. Bella splashes around, lapping up water, observing and corralling the kids, making sure everyone is safe. After an hour, I think we've seen all that needs to be seen of the area, so I ask where we're headed to next.

Korrine says, "Oh, sorry. I thought I told you — we plan to be here all day."

All day? I think. *Hmm. When's the last time we were in one spot all day?*

I sit on a sun-warmed rock, then lie back and look up at the sky and take a deep breath. After a while, Bella hops up and dries off on the rock next to me, just close enough for me to scratch her rump. She's usually ready to go to the next place just as much as I am. Today, Bella seems content to stay right here. I realize that for the first time in a long time we are slowing down to really *be.* I can't remember when I was able to just lay back without planning what to do next. This is an important moment.

This moment matters. I take note of how it feels to relax with the intention of simply doing nothing. Today I have the river and the sunshine and the warm rocks. Friends. Food in my belly. A car that runs. Today I have Bella with me. Maybe that's enough for today.

The next morning, our spirits refreshed from a day of true R&R, Bella and I drive on and hike three miles of the Appalachian

Trail, mostly just to say we did. As usual, we make friends with those we pass on the trail, each inspired by Bella's happy hop. We camp that night and both bathe in a cool river the next morning. After we're out and I'm dressed, along comes one man by himself, so old his beard is completely gray. He tells me he hopes he'll be finished hiking the trail by August.

One man all alone.

No dog. No traveling partner.

I look down at Bella; she looks up at me. I can't help but wonder if that will be me someday, and she seems to know the thoughts running through my mind as she leans into my leg.

Bella and I drive to St. Augustine, Florida, the country's oldest city. It's near the end of March, and already it's summertime here. I receive a call from Adam, a friend in Kansas City, saying he'd like me to help him out with a veterans' program he's running that starts in a week. Bella and I don't have time to drive to the Keys, which are still hundreds of miles away, so St. Augustine is as far south into Florida as we'll get. We turn back north. It's the beginning of April; pink blossoms are emerging on the tree-filled hills when we arrive in Kansas City.

Adam was a medic in a Sapper unit, conducting IED clearance operations during the battle of Sadr City. He survived some of the most hostile fighting of the war in Iraq. Only when he returned to America did he start having problems. Panic attacks. A few years ago, he and I were responding with Team Rubicon to help the tornado-torn town of Baxter Springs, Kansas, and the very day I met him he had an attack. One of the other volunteers quickly stepped in and showed Adam breathing techniques he had learned at a holistic healing program for veterans. It worked better than any drug Adam had ever taken. Adam immediately wanted to attend the program himself. Only one problem: it was in Malibu, California, and flying across the country in a tube full of a hundred-plus strangers to one of the busiest airports in the country, LAX, isn't easy for a guy prone to panic. He asked me to go with him, and I said yes, without hesitation. I had battle buddies when I needed them most, and now this was my time to be there for Adam. In Malibu, we slowed things down through meditation and learned more about ourselves in one week than we had in our entire lives.

Fast-forward to today, and Adam has founded his own nonprofit, bringing the teachings from Malibu to Kansas City. Bella and I stay for the week, and I take photographs to document the journeys of the participants. It's my way of giving back. Bella adopts the role of therapy dog and gives love to anyone who needs it. She seems to mean something particularly special for anyone with a disability of their own — whether their suffering is physical or mental, they can relate. The energy in this program is very open, very vulnerable. That vulnerability, I believe, allows the participants to appreciate the life that Bella has left in her, and knowing that it is finite makes it all the more beautiful.

Through seminars, group discussions, and meditation, we study how to peel back our masks, unpack trauma, and search for our true identities. We learn how warriors have a deep-seated sense of service. We've been trained to serve and fight for others, but when our service is up, we're still looking for something to fight for, a cause to champion. I look at Bella, and the cogs start turning. She's something to fight for. But what do I fight for when she's gone? Seems to me that so many men in the room are just fighting for their own souls, and maybe I'm one of them. In fact, I know I am. This entire journey isn't just for Bella. So much of this has been for me. I'm trying to save myself. I can't save Bella, but perhaps keeping true to my promise to stay with her until her dying breath will help to save me.

With today's battle won, we say our goodbyes and get back on the road toward Nebraska and the little house I've been paying rent on all this time. We've been on the road nearly five months. My first thought is that our grand trip is now complete. We didn't make it as far as we'd said we would, but we took our big adventure, and now it's time to get on with life. Or — and I swallow the thought with difficulty — in Bella's case, get on with . . . *death.*

For a week or two I work for my buddy again, putting door handles on cabinets. Bella is left at home while I'm working. When I get off work, I take Bella for hikes in the park. We are right back where we started. I felt like we were living with purpose out on the road. Now suddenly I'm left wondering if it was all for nothing.

I go home to Broken Bow for a few days and see my family. My nephew Chandler is graduating from high school, and I take his graduation pictures. When I look through the lens, it's as if I'm looking at a ghost. He looks so much like Mike. His facial structure. His sandy-blond hair. His striking blue eyes. Mike and his wife didn't have children, and though that might be considered a blessing, since Mike's death I've often thought that it would be nice for our family to have a piece of Mike to live on. Today, as Mike's blue eyes appear under the brim of Chandler's cowboy hat, I see that we do.

I work for a few days on the farm with Chandler, mainly just tagging along. Chandler tells me one thing he would like to do is travel, but his time off is so limited due to his responsibilities on the farm.

"What if I took you on a trip this summer?" I ask. "Just the two of us and Bella."

"I'd love that," he says.

I smile. "Then it's a promise."

My buddy Kory, comes through town and sees that I'm struggling with what to do next. Bella seems fine right now, but I'm treading water, thinking ahead to when her health takes a turn for the worse, wondering what I'll do then, and what'll happen next. The thoughts run around in my head and paralyze me, and before I know it depressed Rob has returned. So Kory invites me to a swift-water rescue training program in south central Missouri. Bella is invited too, so we go. All ages are at the family-oriented camp, and each day we have some hangout time. Children, teens, and adults all laugh as they splash in the cool water of the river. Some swim, some float on tubes, and the rest seek shade under the large oak trees. Bella hops straight into the mix, swimming after a football that two boys had tossed and missed, wondering if she'll ever get a turn.

I smile as I lie back in the water. The air in my lungs lifts my body afloat. The water runs over my ears, and the calm roar of the moving water is all I can hear. Bella swims over to me to check in, then back to the group of kids. Right now, they're a lot more fun. My sense of hearing subsides again. I take in the bright blue sky above me, and my imagination starts transforming the clouds into faces and animals. The breeze rustles the green leaves of the tall trees that cast large shadows down onto the bank. The sun sends rays of light down onto the river. I remind myself how happy I am to be alive.

My feet sink to the riverbed, and I stand up to make sure Bella's not trying to nose her way into someone's cooler or do something silly. She's a good dog, and as hard as it is to admit, she's still a dog. Sometimes the urge to chase something — a squirrel, a bird, a human she really wants to meet — is just too irresistible. Her friendliness is her biggest attribute and also her biggest vice. I spot Bella in the river about twenty yards from me. She's near a few kids and swimming toward the shore. She's happy, giving a few playful barks. But immediately I see what she can't see. If she keeps swimming in the direction she's headed, she's going to be in danger. Nearby, the current picks up and swirls through four large culvert pipes that carry the stream under the road, each pipe about three feet in diameter. The kids have all been instructed not to go past the crooked tree, but Bella obviously didn't get that memo.

"Bella!" I shout. "Come!"

She doesn't hear me. She keeps swimming, headed for danger.

"Bella — stop!" I'm moving now, trudging forward as fast as I can move through waist-deep water. Underneath me, the river rocks are slick. If Bella doesn't stop, she'll be sucked into the turbulent culvert.

"BELLA!" I scream. She hears my third call, turns, and starts swimming toward me. But it's too late. The current has its hold, and Bella's three legs aren't strong enough to power against the flow.

I move my legs harder, desperate to power faster through the resistant water. Bella looks scared. I've rarely seen fear on Bella's face, as tough and resilient as she is, but she knows she's in trouble. The water is overpowering her. She's frantic. Beginning to panic.

Just as her hind end enters the culvert, I reach her side. I catch her by the collar, brace my bare feet against the twisted rusty edges of the metal pipe, and hold on. The water pummels against Bella's body, making her 80 pounds feel more like 160. How ironic. We're at a camp that teaches swift-water rescue techniques, and we're the ones needing rescue. On the other side of the road, I can just make out some rescue trainees. They're practicing rope throws in the water that rushes out of the other side of the pipes. They hear hollers for help along the bank and are already rushing over. Bella is close enough to me now that I can reach down and lift her under her chest.

"I got you, girl," I say to her. "I got you."

I hold tight with all my strength. I will not let go. One of the trainees comes over the top and grabs her by the collar. Together, we lift her onto the road. She's safe! The trainee extends his hand to me and pulls me out too.

For a few moments I lie completely still on the road. My breath comes heavily, and I can see those same clouds floating in the blue sky. Water pours off Bella's fur like a spigot. She shakes and licks my face. I look her all over, checking for any injuries. She's fine. I squint down at my feet, fearing a giant gash from the rusty pipe. Not a scratch. We're fine.

Bella doesn't seem fazed by the ordeal. She gives me one final lick and hops right back into the water — this time out of harm's way. Her stroke is still strong, and she looks so happy. So relaxed. Even after being tumbled so vigorously, she is still unafraid of water. She still comes back for more.

I have so much still to learn from this dog. Just moments ago, we came close to drowning. But Bella hasn't let the trauma of this experience dampen her enthusiasm for swimming, the activity she loves. Bella's gone back to the fight, back to the experience of living each moment to its fullest. That's been a theme this entire trip. I see how Bella enjoys every second she's on this planet, and her example inspires me to live the same way. How easy it would be to focus on Bella's dying. But we're deliberately not doing that. We're focusing on her living.

We can't stop death. That's inevitable.

But we can change the way we live.

Later that night at camp, as I'm slowly swinging back and forth in my hammock, I look up at the dark sky and exhale. A few stars peek through the pines above me. Bella lies nearby, dozing peacefully, and I use a stick to push against the ground to keep my hammock swinging as my mind replays the day's events. Something good happened today, and I only realize this in the quietness of the evening:

I'm at my best when helping others.

Today I was reminded that when it counts, I act. Without hesitation.

Today in the river, I stepped up when needed. I put another being in front of myself. And this experience with Bella in the river reminds me this is a part of myself I often forget. My mind holds on to my failures, replaying them like bad movies. Yet, if I could see the good in myself, I can hold my head high and walk tall. I've towed more people out of the snow, jumped more car batteries, stopped at more accidents, and helped others in more ways than I can count. I do these deeds not to be seen doing them, but because that is who I am, that is who I was raised to be, and Bella has reminded me that I need to be proud of that.

I hop out of my hammock and walk over to where Bella lies on her blanket. She's zonked out after a physically active day and doesn't hear me approach. I lean over her body and kiss her perfect forehead.

"Thanks for the reminder, baby girl."

We leave camp, but before heading home to Nebraska, I realize we're only a few hours from Kentucky, the only state east of the Missouri River we missed on our trip. *No better time than now.* I take Bella off-roading in the Land Between the Lakes National Recreation Area. My dad used to take me off-roading in the summers when I was a kid. He had this little Mazda 4×4 pickup that was so much fun, and I loved those moments with my dad. Now here I am, taking my fur-kid out on off-roading adventures of our own. Ruthie handles the steep grades like a champ, and Bella hangs her head out the window with wide eyes, dodging tree branches as we navigate the narrow trails. We stop to take a dip in the lake, and I laugh once Bella finds the water to be as warm as bathwater. She looks at me with bewilderment and disappointment, and I try my best to coax her farther into the lake, promising the water gets cooler as it gets deeper. She's not convinced and runs back to Ruthie. I can't help but laugh at the thought of Bella running away from water instead of toward it.

Back on the highway, Ruthie's steering hesitates while turning left. She must have picked up some mud in her undercarriage from the off-road park, but she can still drive down the smooth asphalt with ease. I realize we're not far away from Nashville, and we can't say we've been to Tennessee if we haven't been to Nashville. Besides, I have a good cycling bud from Project Hero who lives there and he's invited us to come down and see the city. I know I need to get back to work soon. But, man, it feels good to be back on the road again. Just like that, Ruthie's steering allows more leeway. We drive into Nashville and meet my bud downtown. Every bar has live music. Every act sounds good enough to have their own record. We stop by the Grand Ole Opry, and Bella says howdy to everyone she meets.

As we're getting ready to leave, my buddy asks me for a ride to Jacksonville, North Carolina. He has enough money to fly or take the bus, but he doesn't know how he's going to get his dog there, a handsome-faced pit bull named Bruce. At first, I tell him I can't swing the trip — I really need to head back to Lincoln, I am just getting settled in back home. Then I take in the bigger picture as I see him and his dog interact. If I don't act now, if I don't give my bud a ride, then he'll be separated from his best friend. I imagine having to leave Bella behind, and it rips my heart in two. In this moment, I have a chance to stop that from happening to him, and I know he'd do the same for me and Bella, so I have to step up. Bella and I drive him and Bruce the two days to Jacksonville.

That's when it dawns on me: *We're back on the journey.*

Bella and I aren't finished. There's more work to be done, and we have just been reminded that what we're doing is bigger than the two of us. I call my landlord and tell him I'm not going to renew my lease. He's fine with this news. He plans to sell the house, and he gives me a deadline of August 1 to get all my stuff out. That gives me one more month to keep going before we have to get back to Nebraska again.

This decision makes me happier than I've been for weeks. Bella is alive and enjoying every moment we're together, and I start thinking of where to go next. We've already covered every state east of the Missouri

River. Besides North Dakota, the northwest is all that remains. Wait. Florida! We barely crossed the border on the first trip! This time Bella and I are going to make it all the way to the Keys. To mile marker zero, the southernmost point in the continental United States. Bella is with me, and I am with Bella. We're a team, and we're doing this together.

10
A Sense of Accomplishment

Soon after Mike's funeral was over, I flew back to Iraq to finish my deployment at Al Taqaddum. Everything felt different after Mike's death. Nothing had changed on the base, but my perspective would never be the same again. At first, it felt like a fog was clouding my thoughts. Then I started to wonder if there had always been a fog, and if it was only now clearing for the first time. Seeing things with this new sense of clarity felt overwhelming. The world I'd always lived in suddenly no longer existed.

My motivation was gone. The fire inside me, fueled by the pride of being a Marine, was suddenly extinguished, leaving a cold emptiness. "What is all of this for?" "What am I? Who am I?" I went on fewer and fewer repair missions around the base. Morning after morning, I kept trying to go to the gym, but my arms didn't want to push the weights. My legs didn't want to propel my body forward during a run. I'd stopped working out in the mornings, and the last box of Chex Mix that I'd received in the mail I'd eaten all myself, not sharing it with my platoon like I usually did. My physical frame deteriorated along with my mental state.

Now that Mike was gone, it dawned on me that much of my motivation to lift — to do anything well — was only so I could report back to Mike about my success. But now, well, what was the point? I found I didn't give a shit about anything anymore. Not even myself. I hurt, I ached, and I grieved — and nothing mattered.

The winter wind blew cold off the desert and a constant grittiness filled the air. I found myself sitting alone in the mechanic's shop after hours, after the platoon had cleared out. The irony wasn't lost on me that "Taqaddum" is an Arabic word for progress, but I wasn't feeling much progress myself. I was miserable. The most miserable I'd ever felt. Dark waves crashed into my consciousness and flooded my thoughts.

I had my M16 rifle in my hands and live ammo, and I thought, *This would stop it.* I took one of the rounds out of my magazine and traced it along my face to feel it up close — imagining myself setting the butt of the rifle on the floor, the barrel under my chin, reaching down to pull the trigger. I ran the bullet over the side of my cheek, my chin, my lips, my forehead. All it'd take was for me to load and rack this round, switch over to fire, and with a slow and steady pull of the trigger, I would cease to exist. All would be quiet. The thought of the silence and calm felt so inviting, so easy, so final.

In the heat of my misery came a moment of clarity. I realized that wanting to stop the agony by ending my life was completely selfish. Look at the distress caused by Mike's death — the wide-reaching repercussions of his sacrifice. How could I take my own life? How could I do that to the people I loved?

Suicide was not the answer.

When you decide to take your own life, it's not only your own life you affect. You throw a boulder into a pond and cause a giant ripple in every direction, strong enough to knock over those who you love most. I had to fight to find my reasons to stick around, but it came down to this: I love and respect my friends and family too much to ever put anyone through that. If you've ever had those thoughts, just imagine your mother, father, spouse, child, siblings . . . anyone who loves you standing over your casket and bawling, their hearts shattered and lives wrecked because of a decision you made. That's what I did. That's why I'm still here.

I put the round back into the magazine, walked outside, climbed up on a Hesco barrier, and looked out into the night sky. In one direction, I saw the glow of a city I thought might be Fallujah; in another, the nearby town of Habbinayah, and perhaps even a tiny glimmer of Ramadi in the distance. I didn't even know where I was, so what was I doing here? I felt so lost, so tired. Just so damned tired.

The leg pain didn't help. At times, I had gotten used to my physical pain, the inexplicable nerve condition that dogged me. After Mike's death, most days, the pain intensified. I wished I could rip my leg out of my hip socket. The pain compounded my depression, and my depression compounded my pain. It was like a slow leak, draining the will from my spirit. I could still hobble a run, still do pull-ups. But our new combat fitness test, complete with fireman carries and ammo box runs, dropped me to my knees. I got put on limited physical training.

They say not all wounds are visible. Well, this mysterious pain was a scratch compared to what other service personnel had been through. Yet I couldn't deny its reality. The pain was there, and the pain was always there, and the pain wasn't visible to anyone except me.

Once I'd been a motivated young marine, respected by my peers, superiors, and subordinates.

Now I felt broken and defeated, a ghost of a man.

The pain was still there when my deployment finished. I returned home from Iraq in March 2008, and all I wanted was to leave the trouble behind and focus on Charli and Bella. A WELCOME HOME banner hung from the rafters in the gym in Omaha, and my family and Charli's family were there. Everybody greeted me with open arms, and my arms were open in return. I was so happy to see everyone. I was truly happy to be home. Yet it wouldn't really feel like home until I was greeted with that thumping tail. My, how Bella had grown so much in the last few months. Her limbs were long, and her shoulders were broad and muscular. "You've turned into a real dog!" I said as she ran toward me and barreled into my legs. Her newfound size was evident as I nearly toppled over. "I told you I'd come back, baby girl! Now, where's that sock? We've got some unfinished business!"

Each marine in our unit stayed on active duty for a few weeks for post-deployment processing, as is normal upon return. My CO ended up keeping me on for longer in a new role of career advisor so I could continue to seek care for my leg pain. I went through endless tests to track the source and doctors attempted nerve blocks to stop the pain, but we couldn't find a solution. I had to come to terms with the fact that chronic pain had become a part of who I was. Finally, the military put me on a med board, a process that can lead to medical retirement — and I was reluctant at first even to consider it. Marines are supposed to get medically retired because they've lost limbs, not because they've hurt themselves by lifting too much or by jumping out of the backs of trucks. I've even met marines who fight to stay in the military after losing limbs, simply because they are that dedicated to the mission, to their brothers and sisters in arms. Yet here I was looking at quitting because of a bit of pain?

But it was far more than that.

Part of the problem was that I wasn't entirely sure I wanted out of the military. If the med board went through, it meant I could never serve again. I wouldn't have the chance to rejoin the brotherhood, to serve a greater purpose. Though it was only the reserve, being a marine was still the foundation of my identity, and to lose that footing felt daunting. There were still opportunities to make a difference. If I overcame this trivial pain and put in a warrant officer's package, as I had been encouraged to do by my command, I could advance my training and bring the experience back to our unit. I could ensure the marines would actually be prepared to do their jobs when called upon. Perhaps I'd save them from having an experience like mine. But could I even handle that? How could I help prepare marines to become better equipped for the mental warfare they'd battle inside if my own mind felt shattered into pieces?

I wanted Mike to be here. To counsel me. To point me in a direction. He couldn't answer me, so that's when I sought the counsel of Brandon Cooley, my commanding officer. I respected his advice enormously. For him to tell me I'd done enough . . . well, that prompted me to listen.

In the end, the med board came through and I signed the papers.

Just like that, I hung up my uniform.

■ ■ ■ ■

It took a while to adjust to my new life. No Marines. No Mike. No motivation. Bella helped. One evening, maybe seven P.M., I was sitting on the couch again, lost in my thoughts as usual. Bella brought me a tennis ball, dropped it at my feet, then backed up, tail wagging and eyes determined. She looked at me. Looked at the ball. Looked at me. Looked at the ball. "Throw the ball. Please, Daddy, throw the ball!"

I threw the ball. Bella retrieved it. I threw the ball. Bella retrieved it. After three hundred throws, it was around ten P.M., and our nonstop game of fetch had nearly worn a path in the carpet. Bella walked slowly over to me. She dropped the ball and sat down, panting, her body finally showing signs of fatigue. Still, she looked at the ball. Looked at me. Looked at the ball. I realized she'd helped lift my spirits. I'd done something other than just sit there, thinking.

"Tomorrow," I said. "We'll play again tomorrow."

The next morning, I got my mountain bike out of the garage and walked Bella alongside it on her leash the two blocks through our neighborhood to the bike path.

She'd done great on walks, working with a loose leash and learning the rule that she walked to the side and not in front. But if she ever saw a squirrel, another dog, a cat, or a person, she was apt to pull in their direction. A quick correction with the leash usually did the trick. This was my first time to try things out on a bike. I foresaw potential catastrophe ahead but sighed and told myself, *Well . . . no risk, no reward, right?*

I started pedaling. Bella trotted alongside me, keeping the leash loose as if she'd done this a million times. As we accelerated, she switched from a trot to a run. Her legs started to stretch farther from underneath her body, and she picked up speed. I pedaled harder, doing my best to match her speed and not pull her forward or drag her back. I wanted to see what she could do on her own. Ten miles per hour flashed on my speedometer. Fifteen. Twenty. The knobby tires on my mountain bike hummed on the concrete. Bella's nails clicked like the gallop of a small horse. Her eyes darted toward the road. A squirrel.

"Leave it!" I said with the slightest pull on the leash to reinforce the command. Bella's eyes focused back on the trail. This run was too important, and I knew a wholehearted sprint at her top speed would be more freeing for her than chasing some random squirrel. Bella pulled forward. "Let's go, Daddy! Try and keep up!" I shifted into a higher gear and cranked harder on the pedals. A grin spread across my face. Twenty-three miles per hour! Bella flew like the wind.

About a block from our home street, Bella slowed. Fifteen, ten, five, and we were trotting again. I stopped on the trail and rubbed Bella's noggin as she looked up at me with a panting tongue. "Good job, baby girl! Daddy's proud of you!" Her smile made it evident that she was completely content in this moment, experiencing a new form of freedom. "Better than four hours of fetch, right?" I asked. She trotted alongside me the two remaining blocks home, then plopped in her baby pool on the back porch. "That's it," I added. "You just relax and recover in there. I'm sure we zapped most of your energy."

What I didn't know is that cold water can act as an instant recharge for a young lab. Bella hopped out of the water, sending water droplets flying as she shook off. She ran and fetched her ball and dropped it at my feet. She looked at me. Looked at her ball.

■ ■ ■ ■

As a reservist, I'd always had another job, except while being deployed. I had worked nearly every day since I was fourteen years old, but being medically *retired* didn't mean I was done working. It did mean, however, that I needed to find my new life, my new identity. I wanted to spend my free moments with Bella and Charli and enjoy the simple things in life. I loved Charli like I'd never loved anyone before. She loved me. Her letters every day during my deployment showed a level of commitment to our relationship that I'd never experienced before. So I took her to the observation tower in Mahoney State Park, a special place for us, and on the top of the tower, at sunset, I got down on one knee and asked her to be my wife. Through loving and hopeful eyes, she said yes.

The wedding, in Charli's parents' backyard, was picture-perfect. Marine buddies in dress blues formed the famous arch with their swords. My father officiated the wedding. For the first time since I was two years old, both sides of my family were in the same place, and no one was harmed in the process.

Charli's entire family welcomed me as if I was one of their own. They were kind and generous. Charli's mom crafted gifts and quilted blankets for loved ones. Charli's dad helped me start my own lawn business. High and Tight Mow and Snow was the name I gave my company. I bought an old Chevy 2500 pickup that I nicknamed "Ol' Red." She was a single cab, four speed, built to work, and I had all the equipment I needed: a John Deere riding mower and a push mower, a weed eater, an edger, rakes, bags, fertilizer, extra lawn seed, and a little trailer to haul it all with.

Spring and summer, I mowed lawns. Fall, I raked leaves. Winter, I cleared snow from driveways and sidewalks. Bella rode in the back of Ol' Red and often talked me into letting her ride shotgun. Before work, she'd grab the truck keys off the coffee table. When I'd open the door, she'd run right out to the truck, ready to work. And by work, I mean sniffing around yards to find whatever treasures she could find.

Bella came to know our customers as well as I did, and life took on a good rhythm. I felt peace and happiness within my new-found family. I had my beautiful wife, my own business to give me purpose, a perfect, tiny little house to return to at night, and our amazing, loving, and adventurous Labrador. Hosting barbecues with friends became the highlight of this slower suburban life.

But I had to mess it up.

During my time in the Marines, I did my best to boost morale, to lighten the mood and fill downtime with laughter. Marines would gather in a circle and ask me to tell funny stories, to talk like Arnold Schwarzenegger or Rocky Balboa, or to impersonate higher-ranking marines. They'd bust a gut laughing and say: "Man, you've got to do something with this talent. Don't waste it! You've got to chase those dreams!" I'd always planned on someday moving out to Hollywood to give acting a shot, but now I wanted to stay here with Charli and make a real home with her and Bella. Yet I couldn't shake the words of Mike's letter: "Follow your dreams and go for it." The lawn business — well, that was solid and I was grateful, but it didn't seem like it was the dream fulfilled. Mike had wanted to see more from me. If I didn't try I'd never know if I could have made it or not. For the rest of my life, I'd be stuck wondering *What if?*

Charli's response summed up why I loved her so much: "Then *go* out there and give it a shot, Rob. If it seems like a fit, Bella and I will move out there with you soon."

Just like that, we changed the course of our lives.

Our initial plan was for me to move to Los Angeles for six months by myself. If things began to click, then Charli would quit her accounting job in Lincoln, put our little house up for rent, load up Bella, and join me in the big city. Los Angeles held unlimited potential for Charli too, and I was excited to get things up and running, although being apart from Charli and Bella proved difficult. Charli and I had been together for a few years, but we'd been married for only three months. I missed her and Bella so much.

At first, I stayed with my buddy, his wife, and their young boy. It was a home away from home. Right away, I enrolled in the intro class at a high-level improv school. The Groundlings was one of the best programs in the country. With no actual performance experience, I felt in over my head among lifelong thespians, but I passed the class and made it to the next stage. I was so excited that things appeared to be going well. This validated my cross-country journey and I knew that I was going to chase this with every ounce of my being. A confidence and sense of purpose came over me. I was no longer following in Mike's footsteps, I was following his last wishes. I didn't have to *make it* as an actor, I just had to keep trying. That was enough. Charli and I didn't wait for the six months to be up. I didn't want to rush her, but she said she'd always wanted to live in a big city and was ready to join me. Arrangements were made. I flew back to Nebraska so Charli and Bella and I could drive out together.

Bella loved riding in the car, although her experience over the long haul had yet to be tested, and we weren't sure how she would handle a multiday, fifteen-hundred-mile trip. I sold my work truck to my dad, knowing I'd rather have a commuter car in L.A., and Charli and I packed our Accord to the gills, saving a comfy spot with blankets for Bella in the back seat. Charli's parents followed behind us in a moving truck. We were off.

Bella seemed to enjoy the long ride. Every so often she peeked out of her back seat to look out the window or catch some passing scent. For the most part she stayed curled up in her little nest. We drove out of Nebraska and on through Colorado, and Bella became accustomed to quick potty breaks on the grass beside mountain gas stations.

In New Mexico, however, Bella looked perplexed to see there were only dusty rocks to pee on. I tried taking her near an assortment of sage brush and other dried bits of desert foliage, but she wasn't having it. With her nose to the ground, she frantically searched for familiarity. "Where's the grass, Daddy? I always pee on grass!"

"I'm sorry, baby girl," I said. "But you're going to have to make do."

My words seemed to help. Bella calmed down enough to pee on rocks. Her face showed signs of relief.

"Yeah, life in California will mean some changes for everybody," I added. "But we'll figure it out, Bella. We all will."

In the bustling city of Los Angeles, it's not easy to find an apartment that allows eighty-pound dogs. After a lot of false starts we finally found a two-bedroom apartment for rent in a complex in Carson, one of the many cities that comprise the greater Los Angeles metroplex. El Cordova in the Park had a dog run in the back of the complex, and many residents kept dogs, although mostly smaller ones. Bella quickly made friends with an affable Staffordshire terrier named Cassius, who we saw every so often, but most days the dog run was empty, and it was like we had our own backyard to run around in together. I liked that we found a place where we could both play off leash.

Bella's size was something of an anomaly in our new neighborhood in the city. Even though she was her usual friendly self to everyone she met, not everybody took to her immediately. When the night watchman first saw her, he gasped and said in his thick Croatian accent: "My Ghod, man. How many pounds is your dog?"

"Eighty," I said. "Don't worry. She's friendly."

He shook his head in disbelief. "Too many pounds! If she want to kill you, you could not stop."

I laughed. "Watch this." I gently opened Bella's mouth with my hands and stuck my face in her mouth. Bella licked me enthusiastically, and I added, "See? I couldn't get her to bite me even if I wanted to."

I never managed to convince that security guard that Bella was just a great big ball of fun, but he was replaced by a younger man who became good friends with the three of us, giving Bella a scratch on the noggin anytime he saw us out walking. Our letter carrier became one of Bella's favorites too. Getting the mail, which used to be a quick walk to the mailbox on the curb, became a trek through the complex to the big square metal grid of locked community mailboxes. Though the inconvenience was trivial, the opportunity for teamwork became evident.

I would grab the mail key, which I had to separate from the car keys so that a certain someone wouldn't get excited about a potential r-i-d-e upon its jingle, and call out: "Ready to go get the mail?" Bella would perk up from her bed with a tilting head, and I'd add: "Where's your collar? Get your collar." Bella would search the living room and quickly find her collar sitting on the plush leather chair. She'd grab it and bring it over, and I'd clasp it around her neck. "Good girl. Now, what about your leash? Find your leash!" I'd give her a bit of guidance, as her leash would be in the spare bedroom. Once she spotted it, she'd dart over and scoop it up. Her tail would be wagging, ready to get to work. I'd hook the leash to her collar and we'd head out the door. Bella would stay right by my side with a loose leash through the complex. Once we reached the big metal mailboxes, Bella would sit and stare at number seventeen. I'd insert the key and give it a twist to open the small door, revealing a collection of envelopes and random advertisements. I'd gather them and hand them to Bella. "Here ya go, girl!" She'd grab the entire lot in her mouth and trot proudly alongside me all the way back to the apartment, where she'd drop them at the front door. "That's Daddy's little helper!" I'd say, as I knelt down and gave her some appreciative ear scratches.

Some days, we met the letter carrier before our box was full. He was a friendly guy and always amazed whenever Bella gently took the mail from him. He said, "You know, the world would be a better place if more people treated their dogs like you do." I didn't feel I treated Bella any differently than how a dog should be treated — with love and respect. Perhaps the bigger message is that the world would become a better place if we all simply gave each other a chance.

We hadn't been in Los Angeles long when we knew we had to get out of the city — if only for a day. We packed sandals, blankets, and towels and headed to Huntington Beach, where we were told we could find the largest off-leash dog beach in the area. The parking lot was full of cars and trucks with bike racks on the back and surf board racks on the top. Families poured out of vans and SUVs, along with countless varieties of dogs. Mostly smaller and medium-sized dogs, but a few bigger ones too. "Here we are, Bella," I said as we unloaded our car. "This feels a lot more like it!" Bella agreed and pulled on her leash with all her might, nearly choking herself with her collar. "Easy, girl," I added. "We'll get there!"

Bella's ears flattened against her head in excitement. Her nose quivered at the sight of water. Waves crashed against the shore, and I wondered how Bella would react to the surf. As we got closer, we saw doggy heaven on earth materialize before our eyes. Shepherds, Newfoundlands, ridgebacks, spaniels, bulldogs, cattle dogs, all sorts of oodles. Just about any other dog I could name could be seen running up and down on this beachfront paradise. Of the dozens of dogs we saw playing in the sand, only three were in the water. Two goldens and one Labrador retriever.

"Well, Bella," I said. "Do you wanna make it an even two on two?"

Bella jumped up and down, pounding her feet on the sand. "Let me go already!"

I looked at Charli and she gave me the go-ahead. I unclasped Bella's leash and she bolted straight toward the water. Her brown coat glistened with a hint of gold as the California sun highlighted her athletic frame. Just as she reached the shoreline, a two-foot wave rolled in and crashed. As if she'd done it a thousand times before, Bella launched herself into the surf. Her chest collided with the wave and the water broke on either side of her shoulders as she kept her head high. Her legs became powerful paddles and she propelled herself forward before the next wave crested.

Just like that, she was swimming in the open ocean. I couldn't help but light up with pride as she swam over to the local ocean dogs and checked in. After Bella introduced herself, she swam back to shore and bent down to get a nice, refreshing gulp of ocean water. "No, no! Bella . . . wait!" I called. But it was too late. A mouthful of salt water slid down the hatch. Her face puckered in a way I'd never seen before. Her nose curled back, revealing her front incisors. Her jowls quivered as if she was about to lose her breakfast. In true Bella form, she wasn't going to let a mouthful of salt water ruin her good time. She ran to us, her tail wagging, and shook water everywhere like a huge sprinkler, eyeing a stuffed canvas ring in my hand that was attached to the end of a rope.

"Hey, Daddy, that toy in your hand, throw it! Throw it as far as you can! This pond is HUUGE!"

I took a few steps for momentum and

hurled the ring into the ocean. Bella gave chase and hurdled a small wave, clearing it completely, splashing into the water on the other side. The toy made a big splash as it landed, and one of the goldens swam over to investigate. Bella spotted him and powered forward in a race, her chest slightly rising out of the water. The golden, with a substantial head start, reached the toy first, but conceded as Bella approached. Bella grasped the toy in her mouth and swam back to shore with her tail in its vertical prideful position.

Once she reached the beach, however, Bella was swarmed by the nonswimming dogs who'd apparently been lying in wait. They might not have had an interest in swimming, but they sure as hell had an interest in the toys that the swimming dogs retrieved from the ocean. Bella did her best to politely dodge the lazy thieves, but a red heeler was too quick and snatched the toy out of Bella's jaws. She didn't put up a fight but came over to me with a look that said, "Daddy, that other dog took my toy."

"Well, Bella girl," I said, "sometimes we have to share what we have, and sometimes we have to hold on to it with all our might."

Bella decided she could share this toy. She ran back to the heeler and the two dogs ran together. The entire beach came alive with pure happiness. Happy dogs, happy people, happy lives. *Yeah. This beach thing,* I thought. *We fit in here. This is where we belong.*

On our way home, we stopped in the city of Long Beach to enjoy some frozen yogurt. I noticed a uniqueness to the people. They rode beach cruisers in the bicycle lane. They lounged on the patios of the endless restaurants. In front of many restaurants, a dog lay in the shade on the patio with its leash looped around a pole. Most dogs were within eyesight of their owners, but I noticed that some people would go into shops while their dog waited outside, tethered to a bike rack. It became evident that there was a total trust not only in the dogs, but in the community as well. Passersby knelt to pet the dogs and take pictures with them, and strangers would even untangle a leg from a leash if needed. Many shop owners had water dishes outside their doors, and the restaurants with the patios offered water for dogs as well. *This is how dogs and humans should live,* I thought. *A cohesive unit, a bonded family, a trusting community.*

One day, having just arrived at Rosie's Dog Beach in Long Beach, I threw Bella's favorite toy into the water. Bella jumped over an incoming wave, landed with a yip and came limping back to us. I checked her paw pads to see if she'd cut herself, but nothing looked wrong. So we let her try to swim it off. She ran back into the water but still didn't put any weight on that leg. We decided to take her home to rest, which seemed to make her more upset than anything. "Daddy, come on. We just got to the beach. And now we're leaving already? I'm good on three legs — let's keep going!"

After resting up for a few hours, Bella was still limping. That evening we decided to take her to an emergency vet. X-rays showed she'd torn her cranial cruciate ligament on her back right knee, which is like a human's anterior cruciate ligament, or ACL. The CCL connects the back of the femur, the bone above the knee, with the front of the tibia, the bone below the knee. It's one of the knee's major stabilizers, and a complete tear requires surgery. We had two options. One might limit her mobility, while the other would allow her to remain her athletic self. The difference: about a thousand dollars. We opted for the better, more expensive surgery, not the most affordable, about $3,500. It was the option the vets would choose for their own pets and it made sense to me to take it. We scheduled it immediately and put it on a CareCredit card to pay it off over time.

After the surgery, Bella needed to don the dreaded cone of shame so she wouldn't lick herself. Healing took six weeks. No real play, just a lot of lounging around the house. We watched movies together, and her fur grew back, and soon enough her leg was like new again.

This was a moment of realization for me. Before this experience, I had never understood how people spent so much on their dogs — my dad had spent huge sums on multiple surgeries for his dog. I had never gotten it. But I did now. When it came to Bella, there was no doubt in my mind I would do whatever it took to give her the best care available.

While my world in Los Angeles was ever expanding, Charli's world was shrinking. Financial security was paramount in her mind, so she took the first job a temp agency set her up with, at an advertisement distribution company in Compton. Friends tried to help Charli with better job offers, but Charli's loyalty to her new boss caused her to say no. She was away from family and best friends for the first time, and that wore on her. The more I found myself loving L.A., the more I saw Charli burdened by it.

Yet Charli was a trouper, another reason why I loved her, and we both gave the new situation our all. We lived in the apartment in Carson for two years, then bought a starter house in Pico Rivera. Buying a house felt like a real commitment to both of us to make it work in Los Angeles. The next part of the plan called for human kids to add to the furry one we already had. We both wanted children, although I asked if we could wait for five years first so that we could both work on our careers. I wanted to have the financial means to give my kids some of the things I'd had to go without as a child. But after year one, Charli couldn't wait any longer. I pressed to wait. She pressed to hurry up. Something that had long been a happy subject became a dividing force. My thinking was that by the time said kiddo was ready for school, hopefully my career would be built up enough that I could afford to get us into a community closer to my work. Living in the suburb of Studio City was high on my list of priorities. I'd had my heart set on a few town houses in the area before we bought our house. But they didn't feel like home to Charli, while the house in Pico Rivera did.

My goal of finding consistent acting work became an ever-present reality. Soon after moving to L.A., I auditioned for a commercial for the Veterans' Affairs, the first audition I'd ever been on. I booked the gig and didn't make much money, but I was honored to put that uniform back on and advocate for veterans on such a large platform. Not long after that I joined the union and regularly found work as an extra to help pay bills, booked another commercial, and worked on several independent projects to gain experience. I got to be a tiny part of some major blockbusters and meet some of my favorite actors. I joined the American Legion Post 43, Hollywood Post, and a group called Veterans in Media and Entertainment. My new friend circle was full of performers and artists of all types, many of whom were veterans themselves looking to chase their own dreams after service. We went to acting workshops and film premieres. I was really in it, and it was surreal. I was never anybody special, but I was doing it. I was making a living in Los Angeles in the film industry. Not a day went by that I didn't look inside my heart and say to Mike: "I went for it, bro. Just like you said." I was the poor kid from rural Nebraska, the kid who'd been told: "Let's be honest, Rob, nothing will ever happen." Yet there I was, finally living my dream. I felt that I was living it for every single kid who'd ever been told they couldn't.

Like traffic in Los Angeles, not everything in my life was running smoothly. The hour-long commute from Pico Rivera to Hollywood began to wear on me. Resentment started to creep in, as it had been Charli who had wanted to live in Pico Rivera. After leaving an acting gig or an audition in Hollywood or Burbank, I'd be sitting on the freeway in bumper-to-bumper traffic, staring at endless taillights. I'd come across the exit for the town house in Studio City, where I'd wanted to live, but another hour of misery still awaited me. I brought that misery home with me when I walked in the door. I wasn't aware of it then, but Charli, who was finishing her own education and working full-time to support my dream, could feel my resentment, and cracks were starting to form in our perfect world.

11
My Entire World

One midsummer day, our air conditioner wouldn't turn on. I called a military buddy, who came to the house in his work van. When we looked at the unit outside, we discovered that some sort of mysterious razor-toothed goblin had chewed right through the wires.

Bella was busy sniffing the tires of the van, and I called to her. She trotted over with a wagging tail and a look of curious innocence. I pointed to the damaged wires and asked with a reprimanding tone: "Who did that?"

Her curiosity turned to concern and her tail slowed its wag.

"Bella, did *you* do that?"

Her head bowed and turned from me in shame, and she walked away, tail fully tucked. I wasn't really upset, but I wanted to make sure she knew I wasn't exactly thrilled about the situation either. Full disclosure: there's something undeniably charming about the look of a guilty dog. To me, it's the acknowledgment, the understanding. "Oh no. I've done something wrong. I hate it when Daddy knows I've done something wrong." There's also something adorable about the denial. "If I don't look at him, maybe the problem will just go away."

Bella flopped down on the side of the porch and avoided eye contact with me. My buddy started his wiring repair, but I couldn't bear Bella's distress any longer. I called her over. She hesitated. Her head still hung low, and I said, "Come on, it's okay." Still, she moved slowly. I got down on one knee and added a bit more playfulness in my tone: "Come here, baby girl. Daddy's not mad." Bella's head picked up. I opened my arms. She gave a little shake and came running toward me, crashing into my chest as if to say, "Me? Really?" She licked my face and rolled onto her back, submitting for some good belly rubs. Her tail returned to a full wag that thumped on the grass.

"There we go, baby girl, all is forgiven."

Though she might have been smart enough to know what she'd done was wrong, the lesson didn't stick. The very next day the air conditioner went out again. I inspected the wires, and sure enough, evidence showed that the same little goblin had chewed through them. This time, I didn't even scold Bella. I called my buddy again, then headed to the supply store to get some chicken wire to cover the air conditioner and remove the temptation.

We all have moments of chewing through the air-conditioning wires. The temptations that sidetrack us, that knock us off course. My wires would take longer to chew through — and although all would be forgiven eventually, the damage I did would prove unrepairable.

In 2010, a catastrophic earthquake ripped apart the tiny Caribbean nation of Haiti. A year after the earthquake, the country was still hurting. My buddy Kory has his degree in international relief and rescue, so he called me up and said he was going to help and that I should come too. I said, "I don't know how I can pay for a ticket." Charli and I were really low on funds. Kory, who's a Seventh-Day Adventist, said, "God will provide." I'm not religious, but I gave it a shot and shared the need on Facebook, and sure enough, my friends came through and raised the money. I bought a plane ticket to Haiti. There, we helped filter water and taught first aid classes and met with kids in schools and passed out toothbrushes and tried to help wherever we could. What I found in Haiti was something I'd lost once I started acting: a sense of service.

It felt great to help others again. I was good at it and wanted to do more. Charli expressed her support, and Bella was okay if I didn't leave for very long. Once back in Los Angeles, I discovered a group of vets who'd formed a disaster-relief organization of their own. They called themselves Team Rubicon, and I signed up as quickly as I could and soon deployed with the team to Marseilles, Illinois, to help with flood-relief efforts. I loved it all. The hard work. The sweat. The team. The mission. Most of all, the people we helped.

A massive tornado struck Moore, Oklahoma, and I was quick to kiss Charli and Bella goodbye, repack my bags, and head back out into the field. Entire neighborhoods were flattened. Giant trees were toppled. Vehicles were twisted hunks of metal. A hospital actually had cars lodged in its upper floors. We worked right in the center of the devastation, our boots on the ground and our hands in rubble. I felt at last like I was serving on the front lines. This was my purpose. These were my people. I was soon appointed to a team leader position. I was finally serving the citizens of our country in ways I'd been searching for.

I figured if I continued to book paying acting gigs, then I could do the volunteer work in my free time. This felt amazing. Like a perfect resolution to all of life's questions. This was how I'd give back. I could do both. My exaltation was short-lived.

A call came from my acting manager.

"Hey, Robby," he said. "We got ourselves a really big audition. It's to play a marine on *True Blood* — the hit TV show. We want you to come in tomorrow."

"Tomorrow?" I said. "I'm in Oklahoma with Team Rubicon. I'll be here for two weeks. There's no way I can make it back by tomorrow."

"What are you talking about, Robby? This is your big break! Get on a plane and fly back."

"There will be more opportunities. This is where I'm meant to be right now."

We talked back and forth. He hung up in a huff, and we parted ways. I had turned down an opportunity he felt was too good to say no to and, in his opinion, that was the wrong decision. For me, it felt good. Honest. Principled. I'd prioritized my volunteering above an important audition.

The volunteer efforts were to help people in need. Surely I'd made the right decision.

But this break with my acting manager would signal the slow death of my fledgling acting career.

It was the beginning of the end of something else, too. Something far more important.

When I came home from Oklahoma to L.A., I was still riding the high of working in the disaster zone. Charli was happy to have me home and planned a day at the coast with Bella and me. Just the three of us. Just our little family. We went to the beach, but my mind was elsewhere. I was still focused on saving the world, and all I could talk about was how I wanted to keep doing more of it.

"What about acting?" Charli asked. "We moved out here for that."

"I can do that too!" I said.

"What about us? Why can't we be enough for you?"

I didn't have an answer for her. Instead, I had a question that I couldn't articulate: *Why can't you want all of this along with me?*

The more involved in relief efforts I became, the more our relationship floundered and we drifted apart. We both tried to salvage things — for our sakes, for Bella's sake. Neither of us wanted to tear our little unit apart. I invited Charli along with me on Team Rubicon deployments, but she worked forty hours a week, was taking online classes full-time, and couldn't just take a week off to volunteer like I could. So I tried to get Charli to join me in my newfound love for cycling with Project Hero. Charli was an avid runner and I knew she could pick up a new sport quickly. For me, cycling was the first activity I could do since the back and leg pain had taken over. I desperately wanted her to join. I asked and asked again. Finally, she relented and rode with me on an epic tour of Italy and France. She rose to the challenge and did an amazing job, impressing me and everyone on the team with each mountain she climbed with determination. The few times she needed help, I pushed to help get her to the top, and although it wasn't easy, it felt like some kind of reconciliation, some kind of teamwork.

Somewhere near the end of that trip, it hit me. Cycling these mountains was a metaphor. I'd been pushing her our entire marriage to do things she didn't really want to do. Charli was a good sport and often went along with me, but in pushing her I'd stopped her from doing the things she really wanted to do. She'd given up her job for me, her family and friends for me, and put having children on hold. We wanted very different things, and I had to accept that.

Then the tension bristling underneath the surface became too hard to hold under. The day after we returned to the States from France, Charli moved back to Nebraska. We told each other the move was open-ended.

"I just don't know how to support you emotionally," Charli said. "You're so all over the place." It was true. I was scattered. In my quest to add "service" to my life's dream, I'd spread myself too thin.

We were amicable about the entire process. We both genuinely wanted the other person to be happy. We said we'd give each other a year to live our lives separately. Perhaps our two roads would lead back to each other. Bella. What about Bella? This was the first time we'd argue throughout the split. We weren't battling to keep her, we were battling not to take her away from the other. After a few rounds of pleading our cases why Bella would be better off with the other person, Charli convinced me that she would be fine in Nebraska around her family, and that I needed Bella's companionship, and the purpose that she provided me. I knew Charli was right, but I also knew the sacrifice she was making, for me.

On the evening Charli left, we hugged each other and said our goodbyes. Bella and I stood on the doorstep and watched Charli drive away, her taillights blending into the busy L.A. traffic. We walked slowly back into the apartment, and I fell to the floor, deep in grief. Like a child squeezes a teddy bear, I buried my face into Bella's fur and bawled, holding her tighter and tighter. Bella licked the tears off my face. What had I done?

I'd chosen to break apart my family.

Weeks later, while trying to make sense of where this new life was headed, I was offered the chance at a paid position with Project Hero in San Diego, working at the Naval Medical Center, helping recently wounded marines and sailors rehabilitate their injuries through cycling — and my first thought was to call Charli. She was excited for me, supportive, genuine.

"If I took the job, would you consider moving to San Diego with me?" I asked. "Things would be different. I'd be focused. I'd have dependable income."

There was a long pause. "I don't know," Charli said at last. "All my family's here. My family is my home."

I was torn. This was a huge life choice. If I said yes to the job, then I was effectively saying a definitive no to my marriage. I needed to think.

I asked if I could take time to respond to the job, and that hesitation cost me the deal. Within the week, the job opening was posted publicly, and I quickly learned the applicants were extremely competitive. I decided to apply anyway. I didn't have the résumé the others had, but I could prove I had what it takes by being there, where it counted. Partly to support my application, I signed up for the Bluegrass Challenge, another week-long challenge ride, this time across the state of Kentucky. I'd prove my value in person, better than any piece of paper could.

Bella and I hit the road in my Mazdaspeed3. The plan was to see my dad, stepmom, and grandmother in Colorado Springs, then end up at my sister Amy's house in Nebraska to spend the Easter holiday before continuing to Kentucky. Everything went according to plan, then at Amy's, on Good Friday, I received a call from my father.

"Granny had a stroke and she's in a coma," Dad told me. "The docs don't think she'll come out of it."

I was heartbroken. I loved my dad's mom so much. She was truly one of the kindest human beings I've ever known. I wanted to drive back to Colorado to see her one final time, but I'd just visited her a few days earlier, and part of me wanted to remember her as she was then. Plus, if I had any chance at landing this job, I needed to continue to Kentucky and do the ride. I couldn't mess up again. Dad and my stepmom, Donna, both said they understood and respected my decision not to come back. I asked Dad to call me if Grandma passed, and I promised that when that happened, I would drive to Colorado immediately. In the meantime, I knew they were there for her. Grandma wasn't alone.

Two days later, on Easter Sunday morning, I received a call from my cousin. She was bawling hysterically and told me I needed to call my dad immediately. She wouldn't say anything further.

Grandma must have passed, I thought. *No need to be hysterical. Grandma was ninety-three. She had a great life.*

I called Dad. After saying hello, he paused for an unusual amount of time, then calmly presented the news: "Bob, it's something else. The kids were on their way to see Granny. They were in a car accident." He paused again, then added: "Charity was killed."

My sister.

She'd just visited me in California.

Along with their children, Charity and Joy had come out to see me. It had been my thirty-first birthday, and I took them down to Venice Beach so I could share this new life with them. It felt so good to have family around. We had talked about acting and what it was like to live in Los Angeles. They had joked with me that when I hit the big time I was going to look great on the red carpet. I'd said any red-carpet event I was invited to, I'd invite them along. We'd all be there — the Kuglers, the Dohenys, Charli and Bella, so many of those I loved most in the world. That was my plan. There'd be so many of us you wouldn't be able to see the red carpet.

Charity had struggled with her weight for years. She said, "Nah. You'd be embarrassed to be seen with me."

I said, "What? No, I wouldn't. You guys are family. I'm proud of you and where I come from!"

Charity, Joy, and Jason had driven nearly three quarters of the way from Iowa to Colorado, where Granny was in the hospital, now, against the odds, out of her coma.

Charity had been behind the wheel when their car collided with a semi. The truck sent them off the road, and their car rolled. Joy and Jason were hospitalized, though doctors had said their injuries were minimal. But Charity didn't make it. I was going to have to bury another sibling. Another parent was going to have to bury a child. Another boulder had been thrown into our pond. More ripples, more pain, for everyone.

I stayed strong on the phone for Dad. I stayed strong as I walked from the living room into the kitchen. I stayed strong as I sat down at the kitchen table, but then I had to tell Amy. My voice cracked when I told her. Tears dripped onto the table.

I was only a few hours away from North Platte, where Joy and Jason were in the hospital, so I drove over as quickly as I could, hugging them hard as soon as I arrived. My stepbrother, Tony, was there too. We'd fallen out of touch over the years, so it was good to reconnect. He showed up when it mattered. The waiting room felt surreal. It didn't make any sense that we were there. And how could Charity be dead when Joy and Jason looked completely unharmed? Physically, at least.

I asked Joy and Jason what they wanted

us to do. Joy still wanted to see Grandma, but after she and Jason talked it over, they both decided they needed to get back to their own kids, and to Charity's kids, Zach and Carianna. Charity was only in her early forties. Young. She had two children, who Jason would later decide to welcome into his home.

So Tony and I drove them back to Iowa and stayed throughout the evening to make sure they had everything they needed. There wasn't much we could do besides simply be there. You just want to be there for family. That's all you can do sometimes. Just be there.

I missed the ride in Kentucky.

I didn't get the job helping vets.

Before going back to California, Charli and I met with a counselor. Maybe there was still a chance to save our marriage. We all agreed that before I could give my full attention to the relationship without regret, I needed to finish out my lease in California and attack my upcoming auditions with everything I had. One last push. We all understood the power of a looming *what if*? So I returned to L.A. This time, I asked Charli if Bella could stay with her. I hated the thought of being away from Bella, but

Charli had a house in Nebraska with a giant backyard, while I had a second-floor apartment in Los Angeles. I wanted Bella to be able to run free rather than be cooped up in an apartment in a concrete jungle while I was gone all the time on auditions. Saying goodbye to Bella felt brutal. I hugged her for a long time, and her eyes were low, her tail motionless and between her legs. She knew.

I had some hard soul-searching to do. I wrote out my priorities on sticky notes and put them on the bathroom mirror: *Helping people. Acting. Bella. Charli. Chasing dreams. Family.*

That last word shouted at me.

My real dream was to have a family of my own. And with Bella and Charli, I'd already had one. The happiest and most content moments I'd ever had were when I had my little lawn business and my little family around me. As I studied those sticky notes I felt a monumental shift begin to overtake me. I wanted to come home, to live with my family, to live the life I already had. But it was too late. I'd made too many mistakes. I'd chewed through that damned air-conditioning wire one too many times. I'd strung Charli along too far, for too long.

She asked for a divorce.

■ ■ ■ ■

I quickly learned that brings grief. It's the death of a life you once had, and it's truly shocking. From my perspective, it felt as if Charli and her entire family had died in a plane crash. You don't divorce one person. You divorce an entire extended family. One day they were there, and then they were gone. And so was Bella.

All I had left were the auditions. A few small ones came and went, but I had a big opportunity on the horizon. I'd booked an audition for *NCIS,* and thought of it as the chance to get my "big break" — but when I showed up my mind was in another place, my heart wasn't in it, and I blew it. I didn't have the heart to go on another audition. None of this make-believe world seemed important anymore. I'd sacrificed my family for a dream — and in the end, they were both gone.

A deep depression overtook me. I spent countless hours bawling in the bathtub, wondering what would become of my life. Why did I ever want to be an actor anyway? Was I stuck chasing the dreams of a young boy who saw a way to escape on the television? Was I simply trying to prove everyone wrong who told me I couldn't? Was I seeking approval from the masses? Suddenly, it all seemed so selfish, so hollow. Above the bathtub was a skylight, and I lay there until my skin wrinkled and the sky above turned dark and starless. My mind turned dark along with it. I lost interest in everything.

I needed to leave California. I had to get away from this city. I traded my Mazdaspeed3 for Ruthie the 4Runner and bought a six-by-ten-foot cargo trailer. I put my motorcycle in it, packed up all my stuff, and signed off on terminating my lease. I no longer had a home and planned to travel up the West Coast, through the parks, on over to Glacier and perhaps back down into Nebraska. Along the way, if any place felt like home, I'd just stay there. I didn't much care anymore.

The day I was set to leave on my trip, I got a phone call. Another damned phone call. Grandma had passed away. I dropped my West Coast plan and drove to Colorado for the memorial service at her church, then on to Nebraska for her graveside service. My grandmother had been truly noble. In her last few weeks in the hospital, her biggest joy in life had simply been to pray for all the staff and doctors who were caring for her. She reminded everyone who came into her room to pray for others who were struggling, for all people in need, even prisoners, for they needed saving as well. She lived her life well to the end, without hypocrisy — and that felt inspiring.

Me? After the graveside, I found myself back in Nebraska with nowhere to go. *Now what?* I no longer knew what I wanted. I'd sacrificed our simple yet beautiful life to chase the dream of acting. I'd given up the dream of acting to fulfill a desire to serve others in disaster zones. I'd abandoned my marriage in hopes of being set free to do all those things. I'd planned to set out on a cross-country adventure, but death had again waylaid my plans.

What did I want? What was I meant to be doing?

I got out a pen and paper and started to write down what actions I could take to become the person I wanted to be.

"Finish what I've already started" sat on top of the list. Before I could do anything else, I needed to tie up the loose ends. First priority: finishing my nearly completed associate degree in fire protection technology.

The fire program was in Lincoln, so I figured I might as well dig in right away. I stopped by the college campus and discovered that if I started that very day, I'd graduate nine months later in June. So I walked to the bookstore, bought the required books and mandatory blue T-shirt, came back to the fire protection building, shaved my beard in the bathroom, put on the shirt, and sat down for class.

I checked off number one on the list.

"Find a place to stay" might have been a better place to start, but hindsight is everything. I looked for temporary housing but came up with nothing. When Charli found out I was back in Lincoln she asked if I would take Bella again. Charli had a full-time job and felt guilty about leaving Bella at home all day. I longed to take Bella back — I missed her like a piece of my soul was gone — but I didn't have a place of my own yet. I was crashing on buddies' couches and in my 4Runner, so I was in no position to take Bella. She needed a proper home. But I was afraid that if I said no, then I might not get the chance again. Put like that, there was only one answer. I took Bella back.

It was the best decision I'd made in a long while.

I was divorced. I was in a town where most of my best friends had moved away, sleeping on friends' couches or in my car, working a job installing door handles on cabinets. I was finishing up school for a degree I wasn't really invested in. I just had to finish — I needed to taste some accomplishment again. If you're a finisher by nature, then it's hard to understand that thought. I'd started so many things and left them unfinished; I had to take some sort of control of my life and complete the ones that left the biggest gaps in my life. That was my headspace, and in that swirl of confusion, grief, loss, and despair, Bella became my entire world.

I had moved into a good friend's house that was under construction. The arrangement was to help with the remodel in exchange for rent. I bought Bella an orthopedic bed and used a cot for myself. Without a kitchen, I cooked on a hot plate in the basement. We didn't have much, but we had everything we needed. Every evening we walked to the park, where Bella loved to go down the slide with a sense of joy that could light up the night. One evening I lay back on my cot, Bella by my side, and wrote down the things for which I was profoundly grateful:

I am thankful for my Bella, because she is my one constant.

I am thankful for her tail that wags when I

walk in the door.

I am thankful for her desire to sleep near me, even if it's on a crowded cot.

I am thankful for her wet nose and long whiskers that tickle me awake every morning.

I am thankful for the smile she wears on her face, and for the countless smiles she puts on the faces of others.

I am thankful that Bella loves me unconditionally — even when I am broke, when I am sad, when I am angry, when I am wrong, and when I have failed.

I am thankful she teaches me how little we need in life to be happy.

I am thankful Bella is my best friend.

It's a few months later, mid-August 2015. I've graduated from college, finishing what I'd started, free to move on to the next phase. But Bella has been diagnosed with cancer, her leg already removed, she has passed the three-month mark — the earliest possibility of her passing in the terminal prognosis handed down by the vet — and the clock is ticking on her life. But I've been mostly stagnant, uncertain, and we have yet to leave on our big road trip.

One afternoon, the phone rings. I've just

taken a call which may send me back in motion.

"Your name is up!" the woman says. "You're in. Come on back to L.A."

She's an administrator at The Groundlings, and this call has brought big news. I've just been admitted to the advanced class. It's highly competitive to reach this stage, and if I make it through this class, then I'm back in the game. Back in a community of performers. Back to chasing dreams.

With this call, I'm so close.

I hang up the phone and feel ecstatic. That night I can hardly sleep. The next day is a flurry of running around, packing suitcases, quitting my hole-drilling job, and lining up a place to stay back in the City of Angels. The city where, against the odds, my old dreams may still be alive.

It's midafternoon and I'm coming down the stairs in a hurry, and Bella stands at the bottom with her tail wagging. Her head tilts to one side and she gives me a wide-eyed look that means only one thing. She's ready for adventure. She's already grabbed my car keys. She senses something big is up, and she's ready to tackle it with me.

Suddenly focused on getting back to the life I'd left in California, my emotions are heightened. My eyes pinpoint with determination. I say: "Not now, Bella," and hurry down the stairs to take a load of things out to the 4Runner. Bella blocks my way, and I need to move around her to descend. I walk out to Ruthie, dump the load off, then come back up the stairs. Bella blocks me again. "Not now, Bella," I repeat, more firmly this time. Yet on the next trip down, with my suitcase in my hand, Bella gets in my way again.

"Move!" I holler, and I nudge her aside with my knee. "I'm trying to do this thing, and you're in my way!"

Bella hangs her head and drops my keys. She hops on her three legs over to the landing and plops down with a big sigh. Her inquisitive eyebrows sink over her previously excited eyes. She may not understand the words I've said, but she understands the tone and the meaning behind them.

I set down my suitcase and sit on the floor next to her.

I look at the suitcases. My boxes. My shirts and pants and shoes and mountain bike all ready to be moved. I pet Bella's brown fur and scratch behind her floppy ears. I move my hands down her back and scratch her rump, just above her tail. I let out a big sigh myself.

I have just told my best friend to get out of my way, so that I can go and live my life.

Haven't I learned anything? Haven't I learned that chasing dreams isn't worth sacrificing those I love the most? Yet here I am, doing it again, focusing on myself. I don't want to make the same mistakes again. Not about Bella. Not about Charli. Not about anyone. This moment right here, right now — this moment matters. It is time for me to make a change. It's more important to become the man I need to be, rather than do the things I want to do. Or think I want to do.

"Bella," I whisper. "You are not in the way of my life. You *are* my life."

She feels the reconnection, and her ears perk up, her tail begins to wag, her smile returns. She picks up my car keys and gives them a friendly jangle.

I sigh again.

I call the woman at The Groundlings and decline the invitation. She says I'll be wait-listed for another year. After that, the opportunity will be gone. It'll be someone else's dream. I hang up and immediately second-guess my decision. *Have I made the right choice?*

I take a long look at Bella, the only perfect thing I've ever seen in my life. The answer to my question was obvious. "Let's go to the park, baby girl. There's a new slide you're going to love."

Partners in crime.

Bella and I spent hours admiring the beautiful graffiti painted on the abandoned buildings in Detroit.

Bella and me on the island of Put-In-Bay.

My brother, Mike, and my mom and me, the day I was promoted to sergeant. It's a memory I'll cherish forever.

One of my favorite photos of Mike, taken less than two months before he was killed. When I look at this photo I am reminded of the man I attempted to mold myself after. He was a leader, a warrior, a husband, a son, an uncle, and a brother.

Bella staking her claim as queen of the Adirondacks.

Snow snout

I don't have many great photos of me in Iraq, but this photo seems to sum it up quite nicely. I'm posing in this broken-down crane while another marine is doing the actual work.

Bella always had her head out of the window while we drove, but sometimes she pushed her limits and snuck her leg out, too, no matter if I approved or not.

Love in Philly.

Every morning we spent at Onslow Beach in Camp Lejeune, North Carolina, Bella and I woke up early and walked along the water to witness the majestic ocean sunrise.

Bella and her new buddy Gabriel in what Gabriel called "Bella's house."

On Tybee Island, Georgia.

At a beach in Florida, Bella made friends with two other Labs. Everyone had their own ball, but of course they all had their eyes set on the same one.

Bella in the vast open spaces of the Dakotas.

In Colorado we formed a close community with not only the veterinary staff, but friends that we made while simply walking down the street. Here, we are saying our goodbyes as Bella and I were packing up to head farther west.

After she passed, I rubbed Bella's ear between my fingers for the last time.

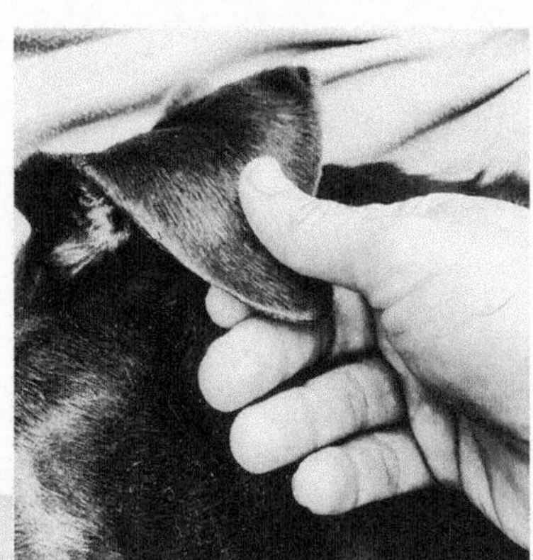

Letting Bella go.

She's always with me.

■ ■ ■ ■

Three: Unconditional Love

■ ■ ■ ■

12
PAWSITIVITY

Bella and I are somewhere near South Carolina, it's late, and we're looking for a rest stop to car camp in, but we're having no luck. We pull off the freeway and soon we're in a residential area — I can't tell where. We see an apartment complex with quite a few open spaces. A decent number of trees. Everything seems secure and quiet. It's not our first choice, but it'll do. We drive Ruthie to the far end of the lot. We're as inconspicuous as we can be. We park, and I take Bella for a quick walk so she can tinkle. In the car, I keep a single thirty-two-ounce wide-mouth Gatorade bottle for the times I can't go in the woods.

Bella and I climb into the back of Ruthie and nestle in for the night. Bella's eyes are already closed, and like turning on a switch, she's already snoring. I'm jealous of that skill. Just before I crawl into my sleeping bag, I strip to butt naked, get out some wipes, and locate the baby powder for my nether regions. For hygiene, on the days I can't shower, I often take a quick pass with a baby wipe or two. After that, a sprinkle of Gold Bond body powder helps me stay fresh enough. Bella doesn't mind what kind of state I'm in, but I'm trying to maintain some standards here. Right at the moment I'm lying on my back with my feet nearly over my head, a knock comes on my window.

"Just a second," I say, and scramble to cover up. It's one thing to be parked where you're not supposed to be. But it's a completely different story when your butt cheeks are in the air.

"What's going on?" says a voice. He's a young man, late twenties, a roving security guard. I'm guessing tenants all have parking passes, and he's glanced at my windshield and seen I don't have one.

"Uh, I'm traveling with my dog," I say through the crack in the window, "and wanted to find a safe place to sleep for the night." Bella is awake now. Her copper eyes stare intensely at the figure through the window.

"Well, no big deal to me," the guard says. "But the morning shift will most likely call to get you towed. You'll want to be out of here before daylight."

I nod, and he nods, and he walks away. Considering the circumstances — a half-naked man, a big dog — I appreciate how cool he's being.

Once he's a safe distance away, I resume powdering my nethers to make sure I've covered all territory. Although I often sleep naked in the bag, I make a mental note to throw on some shorts from now on. The cloud of powder reaches Bella's nose and she lets out a giant sneeze, which takes her by surprise. She gives me a look of annoyance. I can't help but laugh. Oh, Bella girl. Thanks for the reminder not to take myself so seriously. You really are the best friend a guy could ask for. We cuddle up and drift back to sleep.

Summer is upon us and it's hotter than blazes in the South. Bella and I drive to Key West, Florida. It's a million degrees outside, so there's no way we can car camp. We have to find a hotel with air-conditioning. Luckily, we find the Banana Bay Resort, a place that cuts us a good deal.

We hop out of the car and immediately Bella picks up the scent of an iguana. Her eyes go to pinpoints. She pulls the leash out of my hand, chases the lizard up a tree, and shoots me a look of astonishment. "Daddy . . . did you see the scales on that squirrel?" Lizards are the new squirrels, and my smile reaches ear to ear as I watch Bella's excitement.

Bella loves to be on the hunt. All I need to do to grab Bella's attention is ask one open-ended phrase with a question in my voice. "Is that a . . . ?" And the words that follow direct the course of her reaction. "Squirrel" sends her eyes darting up into the trees. "Mole" brings her nose to the ground, and "bug" calls her attention to the lamps in the living room.

Squirrels and moles are just a tease, because she's never come close to catching either. Bugs, on the other hand, are a different story. Particularly houseflies. If I'm not on my game, Bella may first notice the annoying buzzing of a fly. Her attentive look makes me aware of the unwelcome intruder. I hop to my feet and scan the surroundings. Bella inspects windows with feverish breaths in and out of her nose. There is no enemy in sight. I call in my best command voice: "LIVING ROOM CLEAR!"

The buzz reappears. I can't place its location. Bella's head tilts toward the kitchen. "Go, go, go!" I say. We run from the carpet to the linoleum. Within seconds, the target is acquired. It's a big one. Up and down the glass, from window to window, the fly buzzes with complete disregard for its intrusion. Bella knows to wait until the winged invader flies at low altitude. Her best bet for a successful strike is the glass of the back door. We both lie in wait, allowing the enemy to be comfortable enough to make such a mistake. In time, our patience pays off. The moment the arrogant fly beats its wings against the door, Bella goes in for the kill. She's not quick enough. It escapes her grasp and flies back to the top of the kitchen window, seeking asylum behind the blinds. Bella gives chase, but the fly is far out of range. I know what I need to do. I step in toward the target and with a calculated swat of my hand, I send it tumbling in a sputter. It regains its bearings and redirects its flight path toward the window, and *smash,* a furry brown missile pins it to the wall. Bella's whiskered jowls open. Her pink tongue lashes out. The fly is pulled into the gullet of the executioner. The threat is neutralized, the home is secure. "Stand down, Bella girl. Stand down." I give Bella a congratulatory pet on the top of her head and add, "We sure do make a good team, Bella Boo." She licks her chops, heads back to the barracks, and lies down with one eye open, always ready for the next battle.

I'm still laughing, thinking about this. Our hotel unit is right next to the pool, and Bella and I survey the insides for winged intruders, but there are none. We live the high life for two nights with a king-size bed thrown in. *Just my luck with the king-size bed,* I think. *Bella will never want to go car camping again.*

We press on and reach mile marker zero, near the famous southernmost point in the continental United States. A sign reads 90 MILES TO CUBA. A line of people wants to take pictures by the marker, but I'm not about to make Bella stand in sweltering July temperatures and burn her pads on the brick, so we find a little dog beach not too far away where she can cool off, and there I shoot some underwater footage of Bella swimming.

Her entire life, I've wanted to do this, but we've only encountered murky waters. These waters are crystal clear. As I shoot underwater, I see every stroke of her three legs as she paddles. The stroke of her front leg pulls the water down through the center of her chest. Her back legs paddle like normal, and her tail works overtime as a rudder to keep her straight. What a magnificent sight, to see her body adapt in a way that I'm not sure her mind ever needed to process. It's remarkable.

"How does she not swim in circles?" people have often asked. Well, now I know.

When Bella gets out of the water, I towel her off and love on her so much that she looks at me wondering why I'm so excited. This is another moment to cherish. A moment I've visualized for years and am fortunate that Bella has lived long enough for us to see together. I inhale deeply, look back to the ocean, and think: *Wow. Here we are. We are really doing this.*

We are living the dream. I think of Mike's letter again. "Dreams." "Go for it." Maybe that's the lesson to learn here. Dreams can change, but no dream will ever be fulfilled if we don't "go for it." Bella and I, we are going for it.

My third cousin works at a news station in Omaha. I've met him only twice, yet he reaches out and asks if he can pitch a story about our travels to his producer. I'm reluctant to give an interview at first, but then I decide that with so much negativity in the world, perhaps people would like to hear a positive story for a change. As we're driving over the iconic bridges that link the Keys, I give a short Skype interview with the anchor at my cousin's station. My phone is attached to a windshield mount, and I talk as I drive.

The story runs the next day. We pull Ruthie over to watch the spot as it airs. Everything seems straightforward. I talk about Bella's diagnosis. How her leg was amputated. How she was only given three to six months to live, but how she's at fourteen months now. Our pictures are interspersed with the story. The anchor describes the bond between a dog and its owner, how that bond can be so strong, even unbreakable, and I talk a bit about the journey and say how I hope it inspires others to dare to live. "I saw firsthand what happens when you don't do what you want to do in your life," I say simply, "and that's that you don't get to do it."

When I've finished watching the piece, I turn around to the backseat toward Bella. "Good job, baby girl. Let's hope we can inspire some folks." Bella wags her tail at whatever it is I'm happy about. And that feels like that.

The next day, the Associated Press picks up the story — and suddenly it's everywhere. Within four days, the story has spread across America. It's gone to India, the United Kingdom, and New Zealand.

We can hardly believe it. Slideshows of our photos on news sites get millions of hits. We start getting phone calls and emails from all over the world from people who've seen or read the stories and are inspired by our journey. We do our best to respond to all the emails and calls. We're stunned. Amazed by the reach of our adventure and its message. The perseverance shown by Bella is inspiring. The joy with which Bella attacks life is contagious. The love of Bella can even be healing.

We find we're becoming unofficial therapists for thousands of people desperate to talk about their dying dogs. Some people wonder what to do. Others want help to raise money for their vet's bills. Others simply want a listening ear. I want to help. I do all I can. Beyond that, Bella and I simply keep going. We just keep taking the trip, putting our heart and soul into it. Keep taking pictures. Keep meeting people. Keep having conversations. Keep learning about life. Keep living to the fullest every chance we get.

In mid-July, I receive an email:

> Hey man, I saw your story this morning. My wife and I are empty nesters and have

a place down in Sarasota. We have a Lab mix rescue named Indy and a pool that you and Bella are welcome to swim in. Stay a day, if you'd like. We'd love to host you. — Stan and Laurie

We don't know Stan and Laurie, but Bella and I decide we need to meet them. We drive over to their house, and these strangers are instantly welcoming. Bella plays with Indy the rescue Lab, while Stan and Laurie take me out to a Cajun restaurant for dinner. Walking into the restaurant feels like walking into an old jungle saloon. Alligator heads and rattlesnake skins are mounted on the walls, and next to the restaurant lies a river with wild gators lurking in the dark water. The restaurant owners made these little bridges across the river for the squirrels to dart across. When people ask me where's the coolest place I've ever eaten during my travels, this is it.

When we get back to Stan and Laurie's, Bella runs to greet me but changes her direction when she spots a lizard. Stan and I are talking, and suddenly I hear Laurie shout: "Rob! Rob! Bella's in the pond!"

We dash over, and sure enough, Bella has chased the lizard into the pond, which is more like a little lake. Trouble is, there's more than a tiny lizard in the pond. A seven-foot gator is on the opposite side, twirling his tail, licking his chops at the possibility of a supper of fresh doggy meat. It can't end like this!

Laurie is closer than me and quickly runs over and tries to pull Bella out by her collar. She's down on her knees, her arm outstretched when I run forward, scoop Bella up under her belly, and carry her back to the yard. I hold Bella close and shake my head, feeling her heartbeat with my hands. Bella, immune to the chaos, looks up to me as if to say: "You think that gator had a chance with me? Come on, Daddy, have a little faith in a girl. I've made it this far!"

That evening we lounge around and watch the movie *Dolphin Tale.* Bella's head rests on my lap and we're both completely at home in these strangers' house — though of course they feel like lifelong friends now. The movie's based on a true story about a dolphin named Winter whose tail is severely damaged after a tangle with a crab trap. A kid named Sawyer befriends Winter and convinces a vet to make a prosthetic tail so the dolphin can swim again. I'm in awe of this story — at the lengths people go to to help animals — I want to meet people like this. At the end of the movie I see how the aquarium that helped Winter is in Clearwater, Florida, not far away.

The next morning, I call up the Clearwater Marine Aquarium and ask if Bella and I can come and visit. We drive over and get the tour. Bella glimpses Winter in her tank and gives me an inquisitive look, as if to say *What in the world . . . ?* The staff hear Bella's story and everybody loves on Bella for a while. As ever, she loves them right back. We take pictures and capture the moment. Then, as we're getting ready to leave, we meet a young girl in a wheelchair. She asks if she can pet Bella, and I say of course, that's what we're here for. The girl pets Bella for two or three minutes before examining her more closely. She sees Bella's missing a leg, and a huge smile spreads across her face. The girl exclaims, "Hey! She's just like me!" I tell the girl about Bella's diagnosis and her perseverance, and she looks at Bella again and adds, "Giiirl, we have *soooo* much in common!" There's more than happiness in the girl's smile. There's a connection. An understanding. *Hey, I get you. This dog gets me.*

Back in the car, I glance at my phone. Our email inbox is full. Our Facebook messaging system is beyond capacity. Our Instagram message box is filled up. I still can't quite believe how many people want to talk to me about Bella — or about their own dogs and what they're going through. We click through several messages, intent upon answering them later. One message says, "I love your story. It was printed in our newspaper today. In Siberia."

The need for unconditional love knows no borders. The quest for a meaningful life is global. I'm now completely at peace with what we're doing, and I have no desire to be anywhere else, or with anyone else, because there's no doubt in my mind that this is exactly where Bella and I are supposed to be.

Bella and I head back to Stan and Laurie's for a second night. That evening they take us to Siesta Key Beach. The sun's setting and the beach is beautiful with white sand and gently rolling breakers. Fifty yards away, a couple is getting married on the beach. Bella rolls in the sand, and I throw a flying disk for her to chase. She lopes into the water for a swim, grabs the disk, lopes out, and rolls in the sand some more. I'm afraid this chocolate sugar cookie of mine is going to lope over to the wedding to see if she can become an extra bridesmaid, or maybe dry off with a big shake next to the bride, but all Bella wants to do is run in the ocean again, play fetch with the disk, repeat.

Some kids approach and ask if they can throw the disk for Bella, and they throw for several minutes, then scamper down the beach, and along comes a man with a prosthetic leg below the knee. We say hello, and Bella lopes up, and he leans down and strokes Bella's coat. His name is David, and he tells me his leg needed to be amputated after he broke his foot and infection set in to the point of no repair. David pets Bella some more, and Bella looks David full in his face. David smiles, and Bella wags.

The next morning, Bella and I say our goodbyes to Stan and Laurie, the strangers who've become friends, and head out on the road. I check my phone and see I've got a message on Twitter from Betsy Landin, the actress from the *Dolphin Tale* movie. Betsy plays Kat, one of the dolphin specialists at the Clearwater Marine Aquarium. We check out her bio and see that after the movie she joined up with Dr. Juli Goldstein, a veterinarian who also had a part in the second movie, *Dolphin Tale 2.* Together, they've created a podcast called *Dr. Juli's Wild World Podcast with Betsy Landin.*

Betsy's asking if we'd like to be on their show, and better yet, if Dr. Juli can check in on Bella's health for us. Dr. Juli lost a dog to cancer and can relate to our story. Bella and I are on the Gulf Coast side of Florida, and they're located in Vero Beach, on the other coast. The clock is ticking. We need to be back in Nebraska in less than two weeks so I can move my stuff out of the rental. But I do the math and figure there's still time. We message them back, then quickly dart the three hours across the state. We do the podcast together in their studio, and they absolutely love Bella, giving her belly and ear rubs. Then afterward we meet up with their friend Steve, an underwater nature photographer, and we all swim in the ocean together. Being around these people who care so deeply for animals is not only comforting, but serendipitous. Days ago, I said I wanted to meet people like them, and suddenly here we are.

After our swim, Steve grills mouthwatering steak kebabs for us. A steak chunk falls off a kebab stick, and before I can scoop it up, Bella's quick to make it disappear. Too bad she hacks it up soon after. As our bellies digest the food and we chat over glasses of red wine, Dr. Goldstein places her stethoscope on Bella's rib cage, listens, and says: "Her lungs sound pretty healthy."

The kind of cancer Bella has — osteosarcoma metastasized to the lungs — is typically very aggressive, and Dr. Juli can't quite believe that Bella is still living with this type of cancer at the fourteen-month mark. She wonders out loud if Bella was originally misdiagnosed. Bella appears so healthy, so happy. I can see how it's hard to imagine Bella has this form of aggressive cancer. Dr. Juli arranges for us to get a full checkup, complete with X-rays, from a veterinarian friend of hers in Yulee, about two hundred miles away.

The next morning, I head up to Yulee and meet with Dr. Jamie Dunn at the Lofton Creek Animal Clinic. She and her clinic welcome us in like family. As she prepares Bella for more X-rays, she tells me she believes there is no way that Bella could have had this cancer in her lungs for this long.

But it wasn't a misdiagnosis originally. I know that already. We've been through this before. I've actually learned to love the moments when the skepticism on a vet's face is replaced with bewilderment. I'd love for there to be a misdiagnosis, of course I would, but I know for sure what Bella's facing. And I know how incredibly well she's handling it.

Sure enough, when the X-ray results come back from the oncologist, it's osteosarcoma metastasized to the lungs. Dr. Dunn's face goes through the same shift I've seen before as she tries to piece together how Bella is still alive. Even more inexplicable is the comparison between the new X-rays and the ones from a year ago. Without explanation or scientific reason, some of the largest lesions have actually *shrunk.*

"I've never seen anything like this before," Dr. Dunn says, her eyes wide. "I can only assume that your continued adventures are what's keeping her alive. Your exploring, your seeing new things each day — this is giving Bella new reasons for living life. Just keep doing exactly what you're doing."

Her statement fuels my soul. This trip is helping others and helping me, and it might be the very thing that's keeping Bella alive. My resolve is renewed: as long as Bella is able to run and jump and swim and hike and go for rides in the car and do all the things she loves doing — that's what we'll do.

Our mission is *to live.*

We can do that. Together, we're good at that.

Bella and I car camp at a trailhead. We wake up to watch the sunrise, and within a few hours, the beach below us fills up with people participating in a Special Olympics surf camp. A young man walks over to us. He tells me his name is Joe-Joe. He has Down syndrome, and his mother is with him, and she tells me he's a two-time cancer survivor.

While she talks, Joe-Joe loves on Bella, and Bella loves on him right back. His mom describes how much it meant to Joe-Joe to meet therapy dogs who visited the hospitals. Back in California, when Bella was well, I had started the process for her to become an official therapy dog, but then we moved back to Nebraska, and the organization didn't exist there. Then, after Bella's diagnosis, the new organizations I reached out to didn't think Bella could handle therapy work due to her condition. "I'm sorry, it's just not fair to her," they told us.

How many times have I thought about this during our adventures? Bella has done more on three legs than most dogs ever get to do on four, traveling across the country, swimming in rivers in Missouri, hiking mountains in the Adirondacks, hopping happily along a couple miles of the Appalachian trail, swimming in the Great Lakes, the Atlantic Ocean, and the Gulf Coast — all the while providing "therapy" for everyone she meets. Including me. Bella didn't need a certificate.

We just needed to keep doing what we were doing. I love this feeling of purpose, and I'm in awe of how much Bella continues to give me. I turn around and take one long last look up and down the beach at Bella's paw prints in the sand next to my footprints. I take a deep breath as the sun reaches higher into the sky, and I let that breath out.

For a guy who once traced a bullet over his lips, feeling so miserable, feeling so hopeless, I realize I've come a long way. And Bella has been at my side every step.

13
Mike's Sunset

A Toyota dealership hears about Ruthie's steering issues and offers to help us with repairs. They need to keep my vehicle overnight, so I pull out my phone to search on Facebook and see if I have friends in the area. Just then, I get a message on Instagram: "Hey, if you're ever in the Florida Panhandle, I have a house, a pool, and a dog. You've always got a place to stay with us." The woman's name is Heather. She and I have never met or talked, yet her profile is filled with dog pictures, which reassures me, so we message her back, and she says, "Come on over."

We meet Heather and her dog, Moose, a chocolate Lab–pit bull mix, who takes an instant shine to Bella. Moose is full of energy and runs in circles. His tail wags so hard he smacks the wall and splits the sheetrock. Bella loves her new friend back but can hardly keep up with his energy. She runs around with Moose for a while, wrestling him and nipping at his toys, then plops down in her bed. Moose keeps trying to play, but Bella gives him a grunt that says, "That's enough, dude."

Heather and I sit and talk, and she tells me how she volunteers with local dog shelters to rescue dogs and help have them adopted. She's really interested in Bella's story and how well Bella is doing, because Heather has seen so many sick dogs that have been put down due to lack of funding. It's a barrier that too many animals and people face.

While Ruthie's in the shop, I'm using a courtesy car that Bella and I can't sleep in as easily as the 4Runner. Heather has a spare bedroom and offers it to us. The next morning, I receive an update from the mechanic saying it's going to take three days to get the right part. Heather's day job requires a lot of travel, and she's heading out on a trip later this morning, but when she hears of our plight she says: "Here's a spare key to my house. You're more than welcome to stay here. Beers are in the fridge." She adds with a smile, "Just don't steal anything. That's the only rule."

I take the extra key from her and study it in my hand. Someone once told me that you can tell a lot about a person by the way he treats his dog, and I wonder if that's what Heather has seen in me. Or perhaps such generosity is simply embedded in this woman's nature. Less than twelve hours ago, Heather and I were complete strangers, and now she's handing Bella and me the key to her house and literally walking away.

In meeting Heather and people like her, I've developed more hope than I've had in a while. This isn't simply a key to a house. This is a key to her *home.* I feel so rich in friendships — even with strangers. Everywhere on this journey, the love that's been shown to me and Bella has been amazing. If we value our lives by my money, then Bella and I are living next to the poverty line. But if we value our lives by love, then we're the richest man and dog I know.

Ruthie is repaired at last, and we head to the beaches of Destin, which I've been told feel like "walking in sugar." There, Bella is quick to introduce herself to any and all beachgoers. After a romp in the ocean waves, she hops over near a young woman sitting on the sand and shakes, scattering water in every direction. It was always going to happen. I run over and apologize, and the girl just laughs and loves on Bella, instantly making friends. The scene of these two together is beautiful, and I ask if I can take a photo of her with Bella. The woman looks so majestic on the beach. Her black hair blows in the wind and contrasts so perfectly with the white sand and beautiful gray sky in the background. Her father is with her, and as I'm looking at the picture on my camera, he says, "I'm sorry to hear about your family. I saw your story online and read about your brother."

We've only just met this man and his daughter, and already he knows what lies deep below my surface. They know the heart of my story, a large part of what makes me who I am. They're asking how I'm doing — not out of pity, but out of compassion and respect.

The father tells me he served in the navy decades ago, and we talk some more about life's path and the obstacles along the way. I realize he's not sad for me, which I appreciate. He's speaking because he knows what I've been through, and although I don't know his story, it's evident he's seen loss in his life too. As I leave this man and his daughter, it dawns on me that I feel *seen.* Other people, strangers, have opened their hearts wider than their eyes, and they have found relatable things that translate into every language: love, loss, heartache, and passion. They reside within all of us.

Requests for help come pouring in. One man asks if I can help him pay his veterinary bills. A woman says her dog needs a new home, can I help? Another person writes and says his dog needs to be transported from California to Minnesota; he doesn't know what to do. A message comes from South Africa along with pictures of dogs being skinned alive — "Hey, can you help us stop this?"

Bella and I can't champion all the requests. It would take a team of people. But we want to do as much as we can. We sift through and decide to send money to help dogs to get spayed and neutered in Greece. Then we call up an army vet whose German shepherd, Shiner, was shot in the hind legs in Texas, and we talk for a long while. The bullet went through both femurs. The dog survived but needs a series of surgeries. I share his story online, and the community comes together to pay for the majority of his vet bills. Another dog, Rocky, has brain cancer, and we help raise money for his treatments. Another dog, Daisy, has lymphoma, and when I meet the woman she explains she's postponed her wedding and has considered dropping out of nursing school to help pay for the treatments. I share her story online, and her bills are paid within twenty-four hours. The journey seems to manifest a sense of purpose larger than just Bella and me. Yet I must shake the outside noise and keep true to my promise to be there for her, so we continue our journey.

Bella and I push hard through the borders of Alabama, Mississippi, and Louisiana, skirting along the Gulf Coast, and head into Texas, periodically stopping for swim-and-fetch breaks. We stop by the Team Rubicon National Operation Center in Grand Prairie, Texas, and catch up with friends I've served with. They all know Bella, and I'd like them to see her one last time. We can't stay long, the deadline for being out of my rental is nearly upon us. The summer swelters on the interstate. We can see the heat shimmering off the black surface.

Near the Kansas border we stop for a potty break. I want to make the stop as quick as possible so that we can retreat back into the comfort of Ruthie's air-conditioning, but before we load back up, Bella darts to the shoulder and gobbles up something lying next to the road. I can't see what's in her mouth, and I tell her to drop it, but already she's woofed it down. I shake my head and say, "Well, I hope whatever you just ate doesn't make you sick."

We climb back in Ruthie and start driving again. A foul stench suddenly invades my nostrils. Bella's sitting in the passenger seat next to me, and her breath smells like a dirty diaper. I take a closer whiff. That must be what it is, all right — she's eaten poop. Baby poop. I roll down the window so I can breathe, but the hot Oklahoma wind blasts at us like a hairdryer, so I roll the window back up and crank the air conditioner. That's a mistake too. Bella pants *hot* diaper breaths at close range, and the stench from her mouth is nearly unbearable.

I tell her to go in the back and lie down, but Bella turns stubborn on me and keeps her seat. I push on her rump to get her in the back, but she only leans closer to me with her poop breath. The more I push her away, the harder Bella leans back into me. She thinks I'm playing a game.

"Bella," I say. "Get in the back!"

Pant, pant, pant. It feels like a taunt.

"Bella," I say again. "Back!"

Pant, pant, pant.

"Bella!" I shout these next words — *"Get*

in the back and lie down — now!"

Bella's ears lower. She slinks into the back seat, curling into a ball of shame.

The air up front finally clears. I gulp a few deep breaths and stare at the shimmers on the road ahead. Then I look into the rearview mirror. I see my beautiful baby girl curled up in a ball. Her big copper eyes hold such a look of sadness. What does she understand about why I just yelled at her? All she understands is she ate a delicious treat, then Daddy snapped at her with ferocity. Waves of remorse wash over me and my tone changes: "Oh, baby girl. I'm sorry. Daddy's so sorry. C'mere, baby. Let me give you some lovin'."

Bella's tail wags with each soothing word. She stands up and moves forward, pressing closer to me near the back of my neck. I reach back to pet her, and she lays a big old sloppy lick right on my face. *Great. Poop tongue.* But this time I only laugh and say, "Okay, maybe not that much love."

Bella's demeanor has completely changed. She's happy Bella again. She lies back down, and I marvel at her forgiveness. A dog forgives instantly. It's not like with human relationships when a year from now we'd replay this conversation. We'd be driving along somewhere, and Bella would give me that look of disgust, and I'd look at her and say, "What?" And she'd roll her eyes and say, "Oh, nothing." And I'd say, "No, tell me." And she'd say, "Okay — remember that time in Oklahoma when I ate a diaper and was breathing in your face and you — (gasp) — *yelled* at me?" And I'd stutter and say, "But . . . but honey, I told you back then I was sorry." And she'd scrunch her nose up in a snit and say, "Well, I think you should tell me you're sorry all over again. . . ."

Nope. That will never happen. Not with a dog.

Late evening, late in July, I pull into Lincoln. It's my thirty-fourth birthday, and I need to be out of my rental quickly. All the next day I pack up boxes and haul stuff to storage and donate things to charities. Finally, I load the last box. Bella climbs into Ruthie's front seat and we pull away.

Once again, we're without a home.

Once again, we don't have a plan.

But that's okay. We always seem to figure it out.

I take Bella to nearby Holmes Lake, the first lake Bella ever swam in as a tri-pawd, and we go paddleboarding in the early August sun. Charli meets us there so she can see Bella again, and Charli and I catch up. Her hair blows gently in the breeze, and it feels good to see a smile on her face.

She's started to see someone new and says she's really happy. That's important for me to hear, because I sincerely want her to be happy. I ask Charli if she wants Bella to stay with her for a few days so they can reconnect, but Charli says, "One afternoon is enough. I think Bella has *the* life on the road with you. I want to see you both keep going. I want to see you finish what you've started."

Charli's blessing for our continued adventures means the world to me. Bella was her dog once too, and it's important to know I'm not taking Bella away from her.

Charli asks if I've met anyone along my travels. I tell her I've met a few people along the way, but none who've stopped me in my tracks. Maybe I will someday again. Maybe.

She asks if I have any idea where the finish will be to our trip, how I'll know when I've finished what I've started.

"Oregon," I answer. "But I'm not sure what will happen from there."

It seems a little arbitrary, but it's always been the end point I've had in mind. The only question is whether Bella will still be with me. What matters most is that she's by my side through the adventures.

What matters most is loving her the best I can. That means keeping her adventuring — and living well — as long as she's able. For so long I have been searching "out there" for some sort of magic elixir to life, chasing dreams like a dog chasing its tail. But particularly since Bella's diagnosis, more and more I've realized that the elixir isn't out there at all. The elixir is Bella, who's right here beside me.

Bella is a bridge to community and love and forgiveness and healing. And when it comes to finding purpose? Well, I certainly haven't forgotten the words from the vet who last examined Bella:

"These adventures are keeping her alive. Keep doing what you're doing."

"You're going to love Oregon," Charli says with confidence. "I have a feeling you might just stay there for good."

I can understand why she thinks that. Oregon seems to have everything I've grown to love. Coastlines and mountains, forests and hills, lakes and waterfalls. I guess we'll have to figure it out once we arrive.

Charli and I say our goodbyes. Perhaps our last goodbyes for some time to come. "I'm sorry." I say, apologizing for being so selfish in our relationship. "You don't need to be, I'm okay. I don't blame you for anything." The words immediately heal some wounds that have been festering since our split. So often, I've blamed myself for everything and have felt less of the man I was before. With these words, I sense Charli is going to be fine, and it makes me feel that something, at least, has reached a conclusion. At our exchange today I sense closure. I sense a bit of peace that I haven't felt in quite some time. Bella hops right back into Ruthie, then looks deeply into my eyes as we're loading up. She can feel a wound has begun to heal in my heart.

Bella and I head to the farm in the open plains of central Nebraska. I remind my nephew Chandler about the promise I made to him a few months earlier at his high school graduation about taking a trip together before summer ends. Chandler is eighteen, and in a few weeks he's heading to the Nebraska Technical College of Agriculture. He hasn't forgotten about my promise, and neither have I. He has a lot of cool camping gear from trips he's taken with his church youth group — fancy mountain meals, collapsible bowls — and when I mention the trip now he says simply: "I'm ready. Let's go."

In the morning we load up, set for our adventure. As I pet Bella behind the ears, she opens her mouth and pants with her famous smile, and I notice a new little bump on her gums. I snap a picture of it and send it to our vet. She soon responds, saying it's probably just inflammation and to get Bella some oral mouth rinse which, after taking Bella in to be observed by a local vet for a second opinion, I do. Overall, the vet doesn't seem to think it's serious, and Bella doesn't seem to be in any pain, so I'll make sure to keep an eye on the spot to see if it gets any worse. I hope it's not as serious as an abscessed tooth.

For a moment I wonder if Chandler and I should take our trip, but I need to continue the theme of finishing things I've started, following through on promises. Bella means the world to me, and I'd do anything for her, yet I don't want to break a promise to Chandler either. Briefly I'm torn. I've been told twice that it's nothing serious, so then the decision is made and we head off. I've got Bella's back. If she needs help, we'll get it for her along the way.

Chandler has a Ford Ranger pickup of his own that he really wants to drive so he can take his kayak, so we head out on the road with two cars, convoy-style, since I have a car carrier with a rooftop tent and two bicycles on the back of Ruthie. We've brought walkie-talkies so we can stay in touch on the road. Cell phones might have made walkie-talkies obsolete, but it's so much easier just to press one button than it is to phone someone. Besides, it's more fun. And Bella has taught me all about fun. We make up call signs and create characters to go with them.

I say, with a country accent: "White Bandit, this is Black Dog, I'm about fifty meters back on your six. Pulling into the next gravel road. My copilot Bella Dog here needs to take a run in the field over yonder and do her business. You're welcome to join, business or not."

Chandler replies, trucker-style: "That's a big ten-four on the field business. Over."

We head first to Badlands National Park in South Dakota. It's like being on another planet, full of endless dirt crevasses that stretch to the horizon. Overhead is big open blue sky. The earth is multiple shades of brown and red — cocoa, coffee, burnt umber, russet, tan. Its texture is like folded corduroy.

We stop by Mount Rushmore and at the nearby Crazy Horse Memorial, which is still under construction. At Sylvan Lake in

Custer State Park, south of Rushmore, Chandler hops in his kayak and heads out on the water. I have an inflatable paddleboard and I blow that up, push the paddleboard a foot from shore, and slap the top of the board. Bella immediately jumps on and lies down.

Standing behind her, I maneuver the board out toward the middle. Bella is completely relaxed. Sometimes she lifts her head and gazes around, at the shore, the birds overhead. Mostly she just enjoys her floating chariot, completely content. Rain blows in with the breeze and falls gently on us, making it uncharacteristically chilly. Bella shivers, and I bend down and massage her coat. Chandler paddles over to an area near the shore where giant rocks jut into the water, and we lose sight of him as he heads into a cave to take shelter. Bella's ears perk up as he disappears, and she gives me a quizzical look: "Where'd he go, Daddy?"

I paddle over, and Chandler reemerges. Bella quickly settles down when she sees he's safe.

"You got room for two more in that cave?" I shout.

"Barely, but we can squeeze ya in!" he calls back.

We all head into the cave and huddle up. The rain falls, and we laugh at our predicament. Before long, though, it clears away, and we head back to shore to set up a camping kitchen. Chandler and I warm soup while Bella munches her kibble. The sun sets above the plateau, painting the sky in shades of pink and purple, and after dinner Bella runs free. When darkness falls, the three of us bed down in the rooftop tent I've recently added to Ruthie.

Late afternoon, the next day, we drive to Theodore Roosevelt State Park in North Dakota and see a large colony of prairie dog hills. In Nebraska, there are a decent amount of prairie dogs, but they're hunted there, so when you see them they quickly duck back into their holes. Here in the park, the prairie dogs are out lazing in the sun, chilling like they own the place, barking at each other, gossiping about what's going on in the south side of prairie dog town. Meanwhile, off in the distance, I see a coyote lying near a ridgeline, licking his chops. It's his dinnertime, and he's slyly inching forward toward the town square.

"Look, Bella," I say. "That's how all dogs used to live."

Bella hasn't spotted the wild coyote yet. She's still intently watching the unsuspecting prairie dogs with her head as far out the open window as she can get it. I can never tell if she wants to hunt the animals she sees, or just play with them. But her look tells me: "C'mon, Daddy — get closer. I wanna sniff those prairie squirrels!"

We leave prairie dog town and wind through the park. Plateaus on both sides of the road rise to greet us. We're chasing the sunset, trying to see as much as we can before dark. As we crest a hill, we see bison all over the road in front of us. There are more than we can count, and the one closest to us is a large bull, his hindquarters bulging with muscle and his front covered in a woolly rust-colored coat. He's grazing on roadside grass without a care in the world.

Chandler and I slow our vehicles to a crawl. Bella runs from side to side in the back of Ruthie. Head out of one window, and then the other. She's so excited, trying to inhale every floating molecule of aroma from these funny-looking cows. "What are these things? I want to sniff them up close! I want to meet them!" Bella wants to meet every living creature she sees, of course. She does not feel threatened — even by bison. She's confident she can handle herself when needed. She's not lost in thought about the past or future. She's so focused on the very moment we are in. A moment when this hairy cow stinks like nothing she's ever met before.

We traverse through the bison and head up the gulley to the crest of another plateau. We reach the top, and in front of us, maybe fifty yards away, a herd of wild mustangs runs across the road. I'm speechless. I've never seen wild horses before, and the sight of their thundering hooves rustling up dust on this summer evening raises the hairs on my neck. The power of real wilderness is unfolding directly before our eyes. I feel connected to a forgotten world, one where horses roam free, not owned by anyone, not having to come back to a corral.

The horses pass, and the sun is heading below the horizon. I radio Chandler to pull over at the next lookout so we can get a better look at the sky. I pull in first and park. There's a little lull, then as I see him pull in, a lump forms in my throat, and I'm amazed to think that this is the same person I carried on my shoulders when I graduated from boot camp seventeen years earlier. Even before then, when I was fifteen, I was there in the delivery room with my sister and our mom when Chandler was born. My sister didn't have any inhibitions about us being there for the birth. Amy wanted her child to be born into a community of love and protection, a community of family who would look out for her child, a tribe who'd watch him grow and help guide his way. I've known this kid literally since he took his first breath.

Chandler climbs out of his truck. He's tall with long limbs and wiry muscles from his days of working on the farm. I'm so proud of him. Even of his truck — the transmission was shot, so Chandler looked on eBay and found a totaled truck in Kansas that he could use for parts. He drove down to Kansas in the farm truck, loaded up the wrecked truck, brought it home, and swapped transmissions all by himself. He did that when he was sixteen. "All I did was watch a YouTube video on how to do it," Chandler told me, as though it were the simplest thing in the world. "So hopefully I got all the pieces where they needed to go." He's now driven thousands of miles on that truck, and the truck is just one physical representation of the skill set he has. Skills that tell me that this young man is going to succeed.

Chandler, Bella, and I watch the sunset until the sun drops below the horizon. The sky is filled with color — reds and oranges, yellows and purples — and in those last fleeting moments, my mind flashes up an image of Mike. He was always so proud of me. He was proud of the big stuff, of course, like when I was promoted to sergeant. But he was also proud of the little stuff, too, like how I could quickly doodle a cartoon of SpongeBob on a napkin at a restaurant, a simple illustration that could put a smile on the face of our niece.

The memory that flashes at me clearest is of Mike laughing on the back porch of the house that Charli and I once owned. It's at my going-away party. Mike's sitting on the back deck and leaning against the house. His mouth is wide open in laughter, his head is tossed back, his eyes are closed. He's just asked me to do an impersonation of a homeless character called Anton the Bum, who carries around an empty pickle jar to pee in. The irony that I've been carrying around a Gatorade bottle for just this purpose of late is not lost on me. But, unlike Anton, I don't wipe boogers on other people, and I don't talk in a high, nasally, tongue-licking voice. I did the impersonation fully animated on my feet, hunching my shoulders with my fingers ready to snap. Mike belted out a hysterical laugh, stomping his feet on the wooden deck.

"Oh, man!" Deep, ragged breath. "You're so good." Wiping tears. "You've gotta do something with that someday!"

I miss you so much, my brother, I think. *I'll love you forever . . . and guess what — it has just dawned on me — I did do something with that, but it has nothing to do with Hollywood. The important thing was when I did it for my family and made everyone laugh on the porch. I couldn't have hoped for a better audience.*

The sky is growing darker, but brighter colors are in the air.

I turn to Chandler and say, "I wish Mike was here to see the man you're becoming. Because he'd be really proud of you."

Chandler nods but doesn't say anything.

I turn my face upward, and Chandler and I are silent. Hearts open. Both of us watching the sky. Breaking the quietness, I add: "Actually, Mike *can* see you. And Mike *is* proud of you."

Late August, Chandler and I drive home to Nebraska. I help him pack up, and we send him off to college. I am one proud uncle. It's time for Bella and me to get back on the road again. Bella is busy chasing farm cats, and I call her over to Ruthie. She chases one more cat up a tree, then makes her way over and hops right in and curls up on her bed. "I almost had that one, Daddy."

"Next time, baby girl. Next time." I lean in and rub her ears. Bella gives my forehead a lick. I've worked up a sweat from helping pack in the summer heat, so I'm not sure if it's love or salt she's after. I don't care — I'll take all the Bella lovings I can get. I stroke her face with the palm of my hand, and she nestles into me. I hold her close, and she sneaks in another lick. Before I shut the door, I decide to check her mouth.

Rats. The bump on Bella's gums is a little bigger. It's hard to tell how much, but it sure isn't any better.

We head out on the road and stop in at a vet in Sargent, Nebraska. It's been nearly fifteen months since Bella was first diagnosed with cancer. The vet takes a look and does a few quick tests. She says the bump is most likely some kind of infection. She gives me some antibiotics for Bella and a different kind of mouth rinse to try. But the vet is unsure.

"It doesn't look like anything serious, but keep an eye on it," she says. Her friendly demeanor and calm nature are reassuring.

I'll monitor the bump. If it's not getting any better by the end of the seven-day antibiotic treatment, then I'll take her in to another vet right away.

It doesn't look like anything serious.
Let's hope it stays that way.

14
Just One More Ride

Bella and I reach Lakeside, Nebraska, the home of my dad and stepmom. My stepmom's sister and her husband own a little restaurant there called the Lakeside Café. The café is long shut down, but it has an apartment attached to it where my dad and stepmom live. Train tracks run nearby, and the windows of the apartment rattle whenever a train rushes past and sounds its horn.

Bella and I stay in Lakeside several days, and toward the end of the visit, Dad and I are sitting on their porch on an orange double lounge chair. Bella is nearby, as always, rolling on her back in the lush, green grass. We're surrounded by plants that've survived the winter indoors and were brought back outside to bring life to the little old café patio. It's evening, and our time on the porch is limited, because the mosquitoes are about to embark on their nightly flesh raid. I remind myself that this "now" is important — this is something my time with Bella has taught me — so I open up and tell my dad how much I appreciate his support during this journey.

He says, "Well, Bob, I do love you. I'm proud of you. I truly believe you're doing what you're meant to be doing."

These days, he tells me he loves me quite often, and I know how important this is for him to say. His own father didn't tell him those words often, so he wants to make sure I hear them. No parent wants to repeat a cycle of hurt. My dad and I started off close when I was younger, but as I became older, I started to resent his absence during my childhood. As a kid, I saw him every summer, and he taught me a number of things, but most of the time, I basically grew up without a dad. I know now that the complexity of his relationship with my mom didn't allow much room to co-parent, but as a child, it was hard not to feel abandoned. We talk some more, and it's important for me to let him know those days are behind us.

I say, "During these last few years that I've left conventional work and have been chasing my dreams, I've learned who truly believes in me and who doesn't. Who truly supports me and who doesn't. You and

Donna have been in my corner every step of the way. I can see it, and it matters. All we have is today, and hopefully tomorrow. I've learned so much from you, and I am very thankful to have you in my life." He nods, although that's not all I need to tell him, and I'm on a roll. I take a breath and add with sincerity, "Dad, I'm proud of you too. You taught me what it means to be a good son. What you did for Grandma, I'll never forget that."

After Grandma woke up from her coma, all she wanted to do was come home. Dad fought a hefty battle to make it happen. It wasn't easy to arrange, but he persevered, and he brought her home and held her hand while she took her last breath in her own bed. I tell him all that just before the mosquitoes descend. Dad sits there, just taking in my words, but he doesn't have a response other than another nod. He's taking things in too, processing them in his own way.

The next morning, I give Dad and Donna a hug goodbye. Bella and I load up in the 4Runner, and as we're pulling away, Dad comes out to say one more goodbye. I stop, and he leans in through the window, pauses, and stares at me a minute. He blinks and shakes his head and says, "Wow, for the first time I just saw my younger self in your face. I saw myself in you."

This is his response to last night's talk, and it's powerful for me to hear these words. They remind me I'm a creation of his. I exist because of him, and I forget this most days. I say, "Don't worry, Dad, this isn't goodbye forever. I'll see you soon."

This feels like a larger resolution between my dad and me. The years behind us haven't always been good. But today is good. And tomorrow is full of promise.

As Bella and I cross the border from Nebraska into Wyoming, I stop and take one final picture of Bella in Nebraska. There's a sunflower in the picture to capture some of the area's essence, and the yellow of the sunflower adds a bit of color against the bright blue midwestern sky. I don't know when we will return to Nebraska again, but I have a hunch that when I do, Bella might not be with me. It feels as though my time to be a good dad to her might be coming to an end.

We head south into Fort Collins, Colorado. Bella and I stop for a quick coffee downtown on College Avenue. As we're walking back to Ruthie, a young, petite woman reaches down to pet Bella. She tells me her name is

Maria, and when she learns our names and that we've been traveling across the country, Maria looks at me with wide eyes. "Hey — are you that guy? I'm from Argentina and your story has been all over the news in South America. Is this Bella? Wait 'til I tell my mom I met you!"

Maria asks how long we're in town and tells me about the upcoming Tour de Fat, an annual cycling event in Fort Collins that attracts thousands of people. It's scheduled for tomorrow and is something that cannot be missed. Her husband and all their friends will be going, and Maria insists we meet at their place tomorrow at nine A.M. so we can all go together.

That evening, Bella and I head up to Horsetooth Reservoir. The campgrounds are all full, thanks to the tour, so we camp in the gravel parking lot of a trailhead. It's dusk, and the reservoir is nearby, so I get out the paddleboard so Bella can go for a quick swim. We paddle into the middle of the lake, and everything slows down. These are my favorite moments. Bella wears her life jacket, and she perches on the front of my board, one paw dangling in the water. A happy smile spreads across her face, and after a quiet meditation, she stands on the board, then hops off and swims in a contented circle. She's not chasing a fish or a squirrel. She's just swimming for the pure joy of it. Then she paddles back to me and I help her back onto the board.

The next morning at nine, we show up at Maria's apartment and meet her husband, Jon, roommate, Eric, and their two pups, and a whole group of friends comes over. Maybe eighteen in all. Everybody is an adventurous dog lover, and we let all the dogs have a joyous romp in the courtyard. The pups hang out in the apartment while we go see the Tour de Fat. I wish I had a bike trailer so Bella can come with us, but she'll be okay sleeping for an hour or two.

The Tour de Fat is a spectacle. People are dressed as Vikings, fairies, circus performers, and cartoon characters. Live bands play funky-cool music, and everywhere people are drinking local beer. The air feels festive, alive with color, and I'm glad we stayed to see the event. Afterward, Maria and Jon invite Bella and me to stay over at their apartment. I'm happy to be back with Bella, and the day's been such a success I can almost forget the worry I feel over Bella's health.

All Maria and Jon's friends come over to watch a movie with us. Everything's going fine at first, but as we're settling in to watch, someone asks out loud if one of the dogs has farted. Curious, I waft some air my direction to see if I can pick up the scent. Sure enough, I catch a rank smell, but it doesn't smell like any dog fart I've ever known. It's fouler. I crawl over to where Bella's lying, massage her coat, and gingerly pull up her lip to check on the bump on her gums.

The smell is from Bella's mouth.

In the dim light, I can see that the bump is now an open sore, far worse than an abscessed tooth. Her mouth smells like rotting tissue. I hold her close and whisper, "Oh, Bella, I'm so sorry. We'll get this taken care of. I got you, baby girl." She looks at me like nothing has changed, and I try to mask the fear coming over me. I pet her coat. It's late — everything will be closed right now, but I need to get her to a vet immediately.

It just so happens that Colorado State University in Fort Collins has one of the best-rated animal cancer research teaching hospitals in the country. First thing in the morning I call the hospital, but getting in quickly is impossible. I post about the problem on social media, and a friend reaches out and says she works at Four Seasons Veterinary Specialists, a clinic a few miles south in Loveland. Some of the vets there are grads of CSU, she says, and their oral surgeon is amazing.

Bella and I make the drive to Loveland, and the clinic can get us in at once. Bella hops into the clinic like she's been there a million times and is excited to sniff the new scents and meet the new people. I've always been amazed at how well she does at the vet — there's no fear, just excitement to be somewhere new, meet someone new. The crew is so happy to meet her too and instantly fall in love. I give Bella a kiss goodbye as I hand over the leash to the vet tech, and Bella hops right alongside her to the consulting room, tail up and wagging proudly. Bella's sedated and I'm called back to see her while she's on the examining table. A comfortable pad has been placed between her and the table. A pink blanket covers her, keeping her warm.

Unconscious, Bella looks like a different dog. Her lips droop as she lies on her side. The vet pulls back her lip and says he plans to remove an infected tooth as well as a small section of gum around the tooth. He'll biopsy the area and send the sample to the university for testing. He assures me it's a routine procedure. The worst-case scenario, however, is that tests will show that the osteosarcoma that's in her lungs has now spread to her mouth.

When Bella comes out of sedation, she's confused for a bit, then is soon her old self. Again, she shows no signs of pain. All we can do now is wait for the results, which will come back in a few days. In the meantime, she's not behaving like a dog whose cancer has spread.

That evening I take Bella to a park, and she's off and running, rolling on the lawn, showing no signs of slowing down. I'm amazed again at her agility, her resilience, her joy. The next afternoon, I meet up with a good friend, Jordon, a navy vet turned fellow Team Rubicon member. We play kickball with some of Jordon's buddies, and when it's my turn to run the bases, Bella can't stand to wait on the sidelines. She runs right alongside me, and everybody claps and cheers. Nobody is immune to the sight of a happy three-legged dog loping along, ears flapping in the breeze. My next turn at bat I blast one deep and turn on my heels to sprint the bases. I can make this a home run! When I round third and head down the baseline, Bella comes charging directly for me. Before I can reach home plate, Bella tackles me and attacks me with kisses. I crawl over and tag the base. Bella's tail wags as she barks with celebration. Not for the run I scored, but to remind me she still has what it takes to bring me down.

A nurse from the clinic, Lisa, offers me her spare bedroom in Fort Collins, and we stay the weekend at her place. Each morning Bella and I go to the local dog park to play. Afternoons, we head back to Horsetooth for more hiking and paddleboarding. Bella meets other dogs, swims with vigor, even bites on sticks with no apparent pain. She's having the time of her life.

Finally, we get the call with the results.

The vet's all business. "It's bad news," she says. And then there's a little pause. "It's osteosarcoma in her mouth." They need to remove nearly a quarter of her upper jaw — a "partial maxillectomy," she calls it. I research the procedure, and although it sounds invasive, I see it's not as bad as it first sounds. It's the surest way to be rid of the new cancer.

The surgery is set for a few days from now. While we wait, Bella and I drive up through Estes Park, and time seems to slow down. Everywhere we look we see mountains and grandeur, the kind of landscape that makes you want to be in it and live forever. A friend, Syndee, and her chocolate Lab,

Trapper, join up with us, and we take our pups to the local swimming hole in Estes Park. The next day we head up to the Granby Ranch ski resort, where we meet up with a marine buddy of mine who runs the bike shop there for downhill mountain biking in summer. He's got an amazing trail dog, Rosco, and we let the pups play and head out on a short single-track ride through the woods, feeling alive and confident. We can breathe well in these woods. It's mid-September, and the leaves are already turning. Over the next few days, we hike, go paddleboarding, and marvel at the amazing views all around us. Bella runs through tall grass, and plays in the lake, and rides on our paddleboard, enjoying every minute.

The time for the surgery arrives. We head back to the clinic in Loveland. The vet examines Bella but then wrinkles his brow. Things have changed, he says, even over the past few days. The cancer is more aggressive than previously thought, and the growth has now spread over the midline of Bella's upper jaw. A new surgery is needed, a full maxillectomy. It will remove her upper jaw from under her snout to behind the canines. But Bella's case, the vet adds, is complicated. Since the cancer is still in her lungs,

Bella might not even heal from the mouth surgery before she passes.

The new surgery is still a viable option, but it sounds so severe — and even then, it might not do enough. It might not get rid of the mouth cancer. If we don't do it, the cancer is going to eat away at her jaw and it'll painfully rot off. I can't let that happen. I remember, as I think it through, that I've met a few dogs who've had this procedure done. Their noses slope down, and it gives them an underbite. But the dogs are still happy, still wagging their tails, still enjoying life.

I want more options. The surgery seems too radical. But the cancer is attacking Bella's mouth so aggressively, we must do something. Fortunately, I'm able to schedule an emergency consultation at the Colorado State University Veterinary Teaching Hospital to see if there's another way forward.

I rush over, and I can't help but notice the aspen trees along the way, a fire of orange and reds. The entire world feels alive with color. We meet Dr. Amber Wolfe at CSU, who brings in a whole team — specialists and technicians of every kind. Dr. Wolfe mentions the same idea of removing Bella's entire jaw, but she also notes there are two other options, both involving radiation. Before we can move forward with anything, they need to do a full CAT scan, which they're able to do right away. The scan takes several hours, and they tell me it's best to find something to do. So I kiss Bella goodbye, tell her that I'll be back soon, and make my way to the waiting room. There, I find myself loving on every dog and learning from every human. People have full and heavy hearts, and animals have time limits. Everybody wants to talk. Hours go by as I learn the names and stories of each dog and its owner. Once again, I find myself connected. Connected through love and loss.

A vet student finds me and tells me Bella has woken up. She leads me to the consultation room where Dr. Wolfe has the CT results. It's a three-dimensional image of Bella's snout, and we can peer through the images layer by layer. As we go farther into the layers, the cancer begins to show up as bright orange spots, as bright as the aspen leaves outside. A fire is burning in Bella's mouth, but this fire isn't beautiful, it's savage and destructive.

Dr. Wolfe explains that one way forward is simply to provide palliative care. This will just keep Bella as comfortable as possible until she dies. *No. Not yet. We're not there yet.* The other way forward is called stereotactic radiation therapy (SRT), where a superconcentrated laser is focused on the specific areas of Bella's body that have cancer. SRT takes three days to deliver, plus another couple days of recovery.

It's late September, and I stay at the hospital for the entire day, discussing options. The words ricochet around in my head. I'm a visual person; I need to see the options on a board. They point me to a consultation room with a whiteboard. I write a list of pros and cons for each option.

1. Remove Bella's jaw. Pros: We'll get all the cancer in her mouth in one shot. Cons: It's drastic and invasive and might affect her quality of life. Healing time is questionable due to the cancer in her lungs.
2. Palliative care. Pros: Less harmful on Bella's body. Cons: Not designed to remove the cancer. Only to make her comfortable.
3. SRT. Pros: Stronger success rate. Cons: Cost, three highly intensive days of radiation.

■ ■ ■ ■

I stare at the list, rolling around each option in my mind. I share the options with a few friends and consult Charli about it too. I hear that I shouldn't hold on, that if it's time, then I should "let her go." Believe me, I'd love to simply let Bella go — if that were truly the best option for her. But she's not slowing down otherwise. To the contrary, she's still loving life. I've never wanted to see Bella poked and prodded by medical procedures. I swore I wouldn't do this. Yet here we are, faced with this huge decision. This evil growth has escalated everything.

Slowly, I add a fourth and a fifth option:

4. Do nothing. Pros: I can't think of any. Cons: The cancer will continue to eat away Bella's mouth until she is unable to eat. Then she'll die of starvation.
5. Put her down. Pros: I can't think of any. Cons: Bella still has perked ears and a wagging tail. There's still so much life left in her. You can see as much life in her eyes as the day she was brought home as a puppy. It wouldn't be letting her go; it would

> be ending her life. There's simply no way I can take that away from her.

My conclusions are these: I cannot decide to "do nothing." And I will not put her down. Not now. Not yet. Palliative care doesn't seem enough. The SRT will cost seven thousand dollars. The CAT scan we've just had is another fifteen hundred. I weigh these figures in my head too. Not to mention the biopsy and those earlier visits. Against my wishes at first, but very fortunately, a friend has set up a GoFundMe account for Bella and me — and the account will cover it all. For once, money is not an object. All because of strangers. Strangers who wanted to help and be a part of our story. Strangers who became a community, Team Bella. Here, in the moments I must make these decisions, all these people are in the room with me.

I decide on the SRT. It will mean three very intense days, plus several days of recovery after that. But the faster we can provide comfort to Bella's mouth and stop the tumor that's there, the better off Bella will be. Again, the SRT won't save her. But it will stop the cancer from continuing to eat away at her bone at a rapid rate. For the rest of her life — as long as she has left — at least Bella will be more comfortable.

The SRT is set to begin in a few days. Bella is given several different medications to manage any pain. She's not showing any signs of pain, but I decide to take no chances. It doesn't escape me that Bella and I have just been handed three full beautiful days in between.

Three beautiful days to live as fully as we can.

Bella has never been told: "You're dying." Bella's never been told: "You have three legs." She runs forward, same as always, and she lives life positively, as though there are countless tomorrows to embrace with open arms. We drive up to Horse tooth, and Bella splashes in the water with other dogs. She chews on sticks. She lopes about the rocky shore, sniffing everything madly under the clear September sunshine. Dogs are known to mirror the emotions of their human counterparts, so I try to keep my energy up and stay positive. I don't want Bella to see me depressed and down. I don't want her to have to comfort me.

We come back to town for lunch and eat at a noodle place that allows dogs outside on its patio. Four children come by and ask if they can pet Bella, and I say sure. As the kids pet her coat, Bella lifts her head and smiles. All the kids are smiling. Bella is smiling. Her eyes close as one of the kids scratches her noggin. The world has been reduced to the six of us — these four children, me, and Bella — and all the world is happy.

We take a quick trip to Wyoming, meet up with a friend and his black Lab, and go mountain biking. Bella runs along the trail like nothing's wrong. Bella and I camp at the rocky granite outcroppings of Vedauwoo park. She sits in the sun and I try to take her picture, but Bella doesn't want to sit still. I speak softly to her, put my fingers on the sides of her head, and run my hands down through her silky brown ears. She winces slightly, the first wince I've ever noticed, and I realize the abscess in her mouth is starting to cause her pain. "Sorry, babe," I say. "This isn't about getting pictures." This is about our time together.

Today would have been my sister Charity's forty-fourth birthday, and I feel more than ever the need to truly embrace each moment of the day. So Bella and I hike at Vedauwoo, and I let her off the leash. Bella runs ahead of me, and we come to a fork in the trail. Bella stops, then hops ahead on the fork to the right for a few paces, then stops again and looks back at me. I say, "You got it, girl. Lead the way!" and I let her choose the direction we go. This is her adventure to choose. We come to another fork, and again Bella pauses, takes another look at me, and I say "Go ahead!" She hops along, her ears flopping in the air with each step. Again, Bella pauses and looks at me. Again, I tell her to go ahead. Again, she turns and goes as she pleases.

Something draws her nose to the ground. She blasts air out of her nose, causing small clouds of trail dust to form around her nose, then she takes in too many small breaths through her nose to count. I've learned that dogs do this to stir up the scents to really get a picture of what they're after, the way we use our hands to waft the scents of a good-smelling food toward our nose. Without a super nose, I can't smell what she senses on the trail, but my eyes spot a small hole in the ground. It's not a rattlesnake den, but could it be the nest of a ground squirrel, prairie mouse, or maybe just a big ol' bug? Whatever is in there, I know Bella will love hunting for it.

"Is that fun?" I ask Bella. "Where's the fun?"

Bella stops sniffing and lifts her head to

look at me with intense anticipation. "Did you say 'fun'? Oh, Daddy . . . I love fun! Is this the fun? Did I find the fun?" Her eyes dart around the immediate area and back to the hole. She starts to dig with her one front leg — it's adorable and inspiring. Before she digs too far into this unidentified creature's home, I run up the trail a few yards and say, "Aha — it's over here!" Bella gives up on the hole and runs toward me, eyes scanning and nose sniffing, and I add, "It's gotta be around here somewhere . . . maybe over *here*!" I run a few more yards up the trail, and Bella keeps searching intently. We call this game "chasing fun," and it's been a favorite pastime of ours throughout our many adventures.

As much as I love playing our game of fun, I'm still curious about where Bella will lead us if I'm not directing the action. I follow her around a switchback on the trail, which leads us to an open area, revealing a series of additional switchbacks. I've been told the trail is mild to moderate, but I keep a vigilant eye on her three legs. If the trail gets too difficult, we'll turn her around. Yet if Bella wants to keep going, I'll follow behind.

At the top of the hike, Bella finds a little formation in the rocks where water has gathered. She hops right in and splashes about — her own private spa. I sit on the rocks nearby, take off my hat, and lie back. The early autumn sun is so warm. Just as I'm about to drift off, a dripping-wet nose hovers above me and a tongue licks my face. Bella grins, as if to say, "Okay, Daddy. Break's over. Let's get back to the trail." I put my arm around her, pull her close, and say, "I like how you choose to live life off leash, baby girl. You can take the lead anytime."

Bella and I spend our last free night camping in Medicine Bow National Forest. We wake to a glorious sunrise, then pack up and head to the hospital in Fort Collins. She has been so happy these past three days — and days like these are exactly why I want to do the SRT. This dog is alive — it's not her time yet.

Bella is a bit hesitant to leave my side this morning. A fourth-year vet student, Vanessa, takes over her leash. I walk a few steps down the hallway with them and assure Bella that she's with good people. I stop walking, and Bella walks a bit more, then stops and turns back to see if I'm coming.

"You're good, baby," I say. "Go on."

I watch Bella look at me. She knows she's

good. I realize it's me that she's worried about. I take a deep, conscious breath, smile, and add: "I'm good." Bella looks up at Vanessa and wags her tail. The cancer is eating away at her body, but it is not dimming her light. Her body is susceptible to the cancer, but her spirit seems to be immune. The procedure will take some time, and they tell me to go, so I'm at a park reading a book when my phone rings.

"Hey, Rob, can you come back in?" Pause. "There've been some complications."

Fearing the worst, I head back to the hospital for a consultation with Dr. Wolfe. She tells me that over the last couple of days, Bella's tumor has grown even more. The SRT will be too dangerous now. The radiation could blow a hole right through Bella's mouth into her nasal cavity. We're back to square one, and I'm feeling defeated. If we could just get ahead of this thing . . . But it's moving so fast. The only option left is palliative care — maintaining Bella's quality of life for as long as possible. The palliative care will mean one treatment per week over the next six weeks.

Treatments begin right away. We stay with Maria and Jon for a night, then at an Airbnb for a night, then at a mountain cabin for folks in wheelchairs, a perfect spot, as it has miles and miles of wooden trails. But they're closing for the winter, so we can only spend two nights there. We need to find a place to stay so we're not couch surfing night to night; this will be the longest we've stayed anywhere since we started the trip. I get a call from Lisa, the vet from Four Seasons, and she says she has a friend with a historic cottage who rents a room. We can stay there — for free — while Bella gets treatment. Again, Bella opens doors to strangers' hearts and homes.

Our six weeks begin. Mornings, Bella and I head to the park. Afternoons, there's a big backyard at the cottage for Bella to chase squirrels. Evenings, Bella and I snuggle on the couch watching TV.

The veterinarian team stays positive. Bella's responding well to the treatments. They have no way of knowing for sure, but think Bella has at least three months of life left in her. They encourage us to still take the rest of our trip when the six weeks are over. They want to see us finish what we started. They want to see Bella swimming on the iconic Oregon coastline. They want to see us finish. Our plan, after Bella's final treatment, is to head north through Yellowstone, on toward Glacier, over to Olympic National Park in Washington, and finally on to Oregon. There, we'll have done it. We'll have proved that the dog whose human was told she had as little as three months left, the dog who many people said should be put down because she would have such poor quality of life on three legs, the dog who wouldn't have enough energy or strength to be a therapy dog, was worth fighting for. We'll have proved that by embracing what life we have left, rather than fearing and fighting death, we opened the door to endless possibilities. If we focus on fighting death, we can only lose. If we focus on living life, we can only win.

September turns into October. The leaves become a riot of colors. The temperature dips, mornings and evenings. We take a day trip to paddleboard up at the nearby Red Feather Lakes. Bella's head hangs out the rear window of the 4Runner; her jowls flutter and her ears flap in the wind. We pull into an empty parking lot at one of the several bodies of water that make up the lakes. I let Bella out of the back of the 4Runner and grab the duffel bag that holds the inflatable paddleboard. Bella starts exploring the area, hunting for fun, as I lay out the board on the ground near a picnic table. I get out the pump, attach the hose, and start cranking. The board begins to take shape and stiffen, and I look over to see Bella wrestling with a branch. Not a stick, but a branch that drags on the ground as she shakes it with vigor. Her tail is wagging with joy and it's upright with pride as she lets out ferocious growls, really showing this branch who's boss. As much as I love playing fetch with her, I love watching her play on her own even more. There's just something about watching an animal in a state of play, something magical.

The paddleboard is inflated and I carry it to the water and drop it in. Bella hears the splash and comes hopping over, dropping the branch like yesterday's news. I bend down to put Bella's life jacket on her. Summer is leaving us and the fall weather is starting to set in. We're at a slightly higher elevation here at Red Feather Lakes, and we can feel the effects. There is a briskness in the air, and the water feels colder than back at Horsetooth. I take a step into the water, put my right knee onto the paddleboard, and use my left foot to push off the shore. The board rocks slightly from side to side, and Bella is already lying down, completely content, ready for a quiet and comfy ride around the lake. The board steadies as I find my balance with both feet. The gentle breeze barely has an effect on the water, as the surface is almost mirror-like. The biggest disturbance in the water is my paddle, splashing as it hits the water, sending ripples on either side of the board. Behind us, we leave a faint wake that slowly blends back into the surface. Cruising on the paddleboard is exactly what Bella loves. It's one of her most relaxed places on earth. There is a certain calmness about her that I see only when we paddle out on the water. I think it's because she loves being in and around water so much. Lying on the board feels like swimming to her, without the work.

We continue to paddle around the lake, and the sun begins to work its way down to the top of the tall pines. When we're about twenty yards from where we started, Bella hops off and swims toward the branch she had left earlier. I beach the board and carry it onto a picnic table so I can deflate it. Bella is content to wrestle that same branch. Her look tells me she's having fun, but as I finish rolling up the board and start shoving it into its bag, I hear something that catches my attention.

Bella is coughing.

When Bella was first diagnosed with the metastatic cancer in her lungs, I would listen to her lung sounds with the stethoscope I had from my emergency medical technician training. I had no idea what were good or bad lung sounds in a dog, but I wanted to at least try. I spoke with the vet and asked what I should be listening for. "Oh, you likely won't be able to hear anything. The sounds may diminish slightly over time, but it'll be nearly impossible to notice. It's when she starts coughing that you'll know it's getting close to the end."

Bella is still working on destroying this branch, twisting her entire body from side to side as she tries her hardest to break it into pieces. She drops it from her mouth and puts one paw on the branch and starts breaking off some of the smaller branches. As she does this, she emits small coughs as she exhales, as if clearing her throat. Small, faint coughs, but they are coughs. I call her over to me. She comes hopping happily along, ears flopping in the way I've come to fall in love with, giving tiny coughs as she breathes out through her toothy smile. I have her sit and I inspect her mouth to see if by chance there's a small twig or piece of bark that's causing this cough. No. No stick. No bark. It's a cough.

I put my hands around her face, rub her ears with my thumbs, and pull her forehead close to mine. "I can finish packing in the dark," I say. "How about we just sit here and watch this sunset together?"

Bella gives my face a small lick and then goes back to her stick and brings it to me. "Whatever you say, Daddy! Here, you seem kind of down, maybe you should play with this big ol' stick!? Wanna help me destroy it?!"

I chuckle. The sunset can wait. We play with the big ol' stick until we break it into pieces.

Back in Fort Collins, I take Bella into CSU. Each week, we are growing closer to the staff and students. Each week, the building feels more like home. Each week, its inhabitants feel more like family. Bella has come to know and love the students and faculty, hopping right alongside them as they take her into the back for treatment.

The small cough is gone, but it's been replaced with a larger yet less frequent cough — usually three coughs followed by a hack. We have an entire team of vets working with us, residents and students included, so I ask one of the vets, "I was told coughing was the beginning of the end. How much time do you think we have?"

That's got to be one of the worst questions doctors of any kind are asked. I know there are no absolutes. No crystal ball. Yet we put so much weight on their educated guesses.

"Honestly, we can't believe she's been alive and active this long," the vet says. "There's no telling how long she has left. Maybe another month?"

In mid-October, Bella has her last treatment, and spirits are high. The nasty growth has dried up and seems to be stopped in its tracks. Bella's cough has lessened, and the vets are excited, telling us once again to keep moving forward. I'm concerned about what might happen while we're on the road, heading west to Oregon. But the vets assure me they have connections in Portland they can hook me up with. It's okay to go. In fact, they encourage us.

I have dreamed of being out on the open road again. I want Bella to see the massive trees in the Pacific Northwest. I want us to go hiking with my friend and his huskies in the snowy cascades of Washington. I want her to experience the peace of Oregon. I want her to have one last swim in the Pacific Ocean. I purchase two small humidifiers, one that I can plug into a USB port in the car, and the other for inside a room at night. The inside humidifier has a blue light that illuminates the mist as it dances in the air. I place it near Bella's face, ensuring that she's getting the benefit of the added moisture. The top of her nose is developing a dry spot, so I rub a tiny bit of coconut oil on it and then lie down beside her, drifting off to sleep as I watch the dancing mist.

The staff from Four Seasons, our new family, arrive to show their support when I host a farewell gathering in Fort Collins at Odell Brewing Company. There's also a dozen Team Bella friends who meet up with us; I want to say hello before we head back on the road. We take up a collection for one of Lisa's friends, whose dog, Rocky, has brain cancer. People have been so generous to help us, and we want to pay it forward. Bella seems healthy and strong as all the veterinarian staff compliment her enduring, happy spirit.

That evening a quiet man, early fifties, a former marine, spends a lot of time petting Bella. He asks me if he can walk her outside, around the perimeter of the brewery, and I say yes. When he comes back, he sits next to her and keeps petting her. His eyes close. His wife tells me he lost his chocolate Lab, his best friend, not long ago. I'm both touched and proud. People see this Lab fighting cancer and winning. People see this dog defying all the odds and living for so much longer than she was supposed to. Bella helps fill people's hearts with wonderful memories of loved ones they've lost along the way.

We pack up Ruthie and head to Denver, where the next night we're meeting some more new people at a dog-friendly pizza joint. Some of them have driven as far as two hours to come spend time with us. I'm thanked for starting such a beautiful community. I'm caught off guard, not quite sure what she meant. I take a look around the giant patio and see all these people conversing with each other. I see people hugging and laughing. I see people intently listening to heartfelt stories. I see dogs of all shapes, sizes, and ages running around. And, of course, I see Bella. She's in the middle of the action, channeling her inner pup, wrestling with other dogs, getting butt scratches from everyone willing, and pandering to anyone who she can talk out of some pizza crust. I don't condone the begging behavior, but I let it slip this once. Yeah. A community. I can see it now. Team Bella. It's an awesome little community.

That night, we stay at a friend's apartment, and Bella and I cuddle up for a good night's rest. As I close my eyes I think about how amazing these past two nights have been. I'm not sure how much time Bella has left, but we're going to continue to try and make a difference together every day. I squeeze Bella tight and whisper, "We're doing it, girl. We're still doing it."

But when I wake up in the morning, Bella's not on the bed. She's curled up on the couch in the living room, and there's a fresh pool of diarrhea near the back door. "Oh, baby girl. You should've woken me up," I say, as I take note of how close she was to the door. Never, in her entire life, has Bella had this kind of accident inside. I can only imagine how badly she wanted to go outside; I'm sure she held it as long as she could. I just wish she'd woken me up. I take Bella outside, and she has a bit more diarrhea. We come back inside and I clean up the mess.

We decide to spend our day at nearby Washington Park before we start the journey westward. I want to see how Bella's stomach settles before we get back on the road for a long drive. We're adventuring a little, checking out the sights and sounds, when I notice Bella has a bit of a shiver, which I attribute to her upset stomach and hope it's not more serious. Eventually Bella perks up and plays in the shallows along the shoreline. She seems well enough, so Bella and I decide to stay to enjoy the beautiful fall Denver day. I've recently purchased a Burley bike trailer, so Bella hops right in the trailer, and we head off for a ride around the lake.

We take one lap, and Bella looks so happy. Her ears flap in the wind. She's not shivering anymore, and her smile is bright and clear. We take another lap, and then another. The sun is out, and each person we pass smiles at us. A guy we pass calls out, "That's so awesome you do that for your dog, man."

"Of course! She's my baby girl!" I call back.

Bella and I take another lap, and another. I want this ride to last forever. I think Bella does too.

Just one more ride.

15
Funny How We Show Our Love

When our last ride in the park is finished, I pull over to a spot near the lake and lay out my Woobie, a tricolor camouflage military poncho liner that most any Iraq/Afghanistan war–era vet can appreciate. For a while, Bella is occupied chasing some geese, then gives up and lies down in the grass.

I ask her to come and lie next to me on the slick Woobie. She walks over, but I notice she's shivering again. This time, it's worse. I'm still hoping her shivering is being caused by an upset stomach, but my heart begins to suspect the worst. *Is this it?* I pet her glistening brown belly as it shakes rapidly, scoot closer to her and sit cross-legged, then lay her head in my lap. As I scratch the ear of the best friend I've ever had, I can't deny that this may be the beginning of the end. This nasty disease that she's fought off for so long may finally be overtaking her lungs. Bella can inhale okay, but her breaths outward seem forced and labored, to the point of causing the shiver. Since her original diagnosis, Bella has lived eighteen months. She wasn't expected to live much past three months. She's my miracle dog, my adventure partner. It feels like now this miracle, this adventure, is drawing to an end.

I look into Bella's copper eyes and, for the first time, I feel her discomfort. I need her to feel my love for her in this moment, so I open myself to the experience. The air smells fresh as I breathe in through my nose. The breeze runs across the bare skin of my arms and touches the back of my neck. I look around the park. It's a perfect fall day being enjoyed by people and animals alike. The sun hangs low, about thirty minutes until sunset. Its glow reflects off the lake as the ducks practice their synchronized swimming. They leave sharp lines in the water that spread out behind them in a V shape. The trees still hold on to their golden leaves. All around me, shades of yellow and orange shake in the calm breeze. What a scene. What a *perfect* scene.

Over this past year, I've imagined the "perfect way" for Bella to go. Although I've tried not to focus on death, only on living life, I've still hoped for something just like this. Bella has always loved to swim, and if she dies beside water — letting her cross the bridge to the other side by her favorite element — then it just seems right. Cradling the sides of her head with the cups of my hands, I gently rub her face with my thumbs.

"You can go," I whisper. "It's okay, baby girl. You've done enough. Don't hold on for me. I'll be okay. You can go."

Bella breaks my gaze and spies a flock of geese roaming in a nearby open grassy area. She hops up on all three legs and lopes toward them. They see her and begin to waddle away; a few take flight as Bella picks up speed.

I guess she's not ready to cross over. Not yet. I get up and run with her, and we chase the geese until they scurry.

The scent of a little creek that runs parallel to the lake catches Bella's nose. She breaks away from the geese and hops down an easy embankment that leads down to the creek. She walks right into the water and just stands there, enjoying the water flowing softly against her body. She laps up a few mouthfuls, and I stand on the shore and watch. Whatever she wants to do, for as long as she wants to do it — I will be there for her. That's the only plan now.

Bella hops out of the water and leads me over to where the bike and the Burley trailer are parked. She sniffs at the bike but shows no interest in getting back in the trailer. Instead, she lies down and pants and I see her give another little wince. The sun dips below the horizon. Daylight begins to slip away. Her discomfort seems to be growing.

A man walks by and stops to talk to us. He's had dogs that have passed. The look of sorrow on his face says it all.

"I'll most likely be taking her to the vet soon," I say.

He nods. "Yeah. Probably a good idea."

He can sense my conflict. If Bella's going to die, then it should be here in the park, by the water. But if she's suffering, then I want to get her help from a vet. I've been with my girl so long, I know her so well. I know her fighting spirit, and I know how tough she is. For the first time, I can see she's begun to struggle.

Ruthie is parked about forty yards away. I'm willing to carry Bella, but I know that if she's capable, she'd rather move on her own. "You got this, girl?" I ask. Bella tilts her head with a look of confidence, rises to her feet, and starts running toward the parking lot. I trot alongside her and watch those flapping ears that I have grown to love so much. The flap came with the hop. The hop came with the amputated leg. The amputated leg came with the cancer. Funny how this by-product of the very thing that's killing her has become so adorable.

Bella lies down in the grass near Ruthie while I quickly disassemble the Burley and load the bike. A woman walking along the trail stops and asks if she can pet Bella. "Please do," I say, and explain what's going on. She kneels and pets Bella's soft fur and tells me she's lost more than one dog over the years, and it never gets easier. I walk to the back of the 4Runner to load up the gear, and when I return to the side of the vehicle, the woman is lying alongside Bella, giving her such motherly love that I don't want to interrupt either of them. I creep up slowly and sit down with the woman. We talk more about our furry loved ones and how much they mean to us, and she gets up and says goodbye with a long hug before walking on. She doesn't tell me her name, and I don't ask, but I will never forget her act of kindness.

I load Bella up and into her bed in the back of Ruthie. She lets out a little groan. Earlier, we'd planned to meet my buddy Jordon and his family for dinner up in Thornton, a northern suburb of Denver.

But I call Jordon and say, "Bella's having trouble breathing. We're headed up to the ER vet in Loveland. It doesn't look good." It's the first time I've spoken about the situation to someone I know, who knows Bella. Jordon replies softly, "Let me know if you need anything, brother."

I haven't eaten anything since breakfast, and Bella hasn't been interested in her food most of the day. So I put "Wendy's" into the GPS and swing into a drive-through to order a Jr. Bacon Cheeseburger with a plain chicken patty on the side. When we pull up to the window I ask the attendant if he likes dogs, something I always do before I roll down the window to let Bella solicit for treats. The Wendy's worker replies yes, so I roll down the rear window. Bella knows what this means, and before the window is all the way open, she has her head out and her nose as close to the inside of the drive-through hatch as possible.

Two employees take turns petting Bella's head while I tell them a bit of her story. One of them hands me our order, and I drive forward. Before we pull out of the parking lot, I stop and open the container with the extra patty. It's hamburger, not chicken. I'd prefer not to have all the grease, but what does it matter anymore? If time really is running short, this indulgence can be forgiven. I blow on the patty to cool it off, then reach back to offer her a bit. Bella woofs it down with barely a chew, as if to ask, "How have you been keeping this from me?!" She's never had the luxury of a greasy fast food burger before. I'm happy I can give her one, guilt free. I'm happy she still has an appetite. I'm happy . . . I can help her enjoy what time she has left.

I call my friend Lisa, the ER nurse at the clinic in Loveland. I so desperately hope that she'll be working tonight so Bella will have people she knows around her. Lisa doesn't pick up, so I leave a voice mail saying Bella's having problems breathing and we're on our way. We hop on the freeway and I reach back to pet Bella and make sure she's still breathing, reassuring her that I'm nearby. Loving her. Lisa calls back and says she's on her way in to the clinic. She'll meet us there. Sure enough, when we pull into the dark parking lot of the emergency hospital, Lisa's just getting out of her vehicle with her dog, Jax.

Bella hops out of the 4Runner on her own. She's still shivering, still coughing, still struggling to breathe, yet she lopes happily through the front doors when Lisa unlocks them. The doors are locked, I realize, because we're here after hours. Damn. This really is an emergency.

An entire team is waiting for us in the back of the clinic, and I'm not sure if they invite me to join them, or if they just know I'm not going to leave Bella's side for a second. Again, I cling to the hope that this might all be from pizza the night before. Maybe it's simply denial, but I can't shake off that tiny thread of hope. We take her to the X-ray room, and I pick Bella up and lay her on the table. I have to leave during the X-ray itself, but afterward I'm right back to help her down. We go back to the regular exam room and I lay Bella down on a blanket on the floor. We attach an oxygen sensor to her tongue and it reads 92 percent.

"Well, ninety-two isn't horrible," Lisa says. "But it isn't good either. I like to see no less than ninety-five percent."

I lie down next to Bella and pet her. The X-rays are back, and they're pretty bad, Lisa says. Bella's lungs are full of masses, full of fluid. She simply can't ingest enough oxygen.

Then I hear the words that no pet owner ever wants to hear.

"I'm sorry, but you need to make a decision."

Make a decision. I roll the words around

my mind, considering the implications. The decision to end my best friend's life. It's me who's going to have to end the discomfort of Bella's breathing difficulties. I'm going to have to end the pain of her bone cancer. But in doing so, I'll also be ending Bella's *life.* And I can do this with a mere nod of my head. The stroke of my pen. I look down at her, pet her soft brown fur, and try to prepare myself for what I know must come, but I hadn't realized that it would come so suddenly, even though I've known about it for a year and a half.

I don't want to put Bella down on a cold, hard table in a vet's office. I've always been adamant about that — ever since I heard her initial diagnosis. I've always wanted it to happen outside and preferably near water. If it were to happen in Nebraska, I would've used Holmes Lake. Here in Colorado, my first choice is the little river hideaway in Lee Martinez Park, a favorite spot of Bella's. The clinic doesn't offer off-site euthanizing, but I have a couple of options. The first is with Priscilla, the holistic vet in Fort Collins who's been providing acupuncture and massage for Bella. I call Priscilla and leave a message letting her know that I'd like her to help us with this process, because she knows Bella so well. The second is a service called

Home to Heaven, which several people in town have recommended. Lisa calls Home to Heaven for me and we learn that the first time they could help is nine o'clock tomorrow morning. We make the plan to wait until morning.

Lisa helps arrange the comfort room for Bella and me to stay in until morning. The staff are doing all that they can to give us the best possible accommodation. I bring Bella's bed inside, her handmade blanket, and my own sleeping bag.

To keep Bella comfortable, she's put on oxygen. She refuses a nasal cannula, a plastic tube that's inserted in the nose. Blow-by oxygen will have to do. Lisa hooks up an oxygen tube to a port built into the wall, and I lie next to Bella and hold the tube in front of her face. Lisa has her dog, Jax, in the room as well, and he lies down next to Bella and me. Even when Lisa leaves the room, Jax decides to stay with us. Lisa and Jax have such a strong connection, and he usually won't leave her side. She's baffled by his decision to stay, and the power of this moment is palpable. Dogs have such a keen awareness of where love is needed most. Right now, Bella needs it. I need it too. So Jax stays where his love is needed.

I ask Bella if she wants to go outside, and

the word *outside* brings her to her feet. I take her to go potty, and when we return, her gums have changed colors due to being away from the oxygen, evidence that her lungs aren't sufficient on their own. Bella lies down, and I lie down with her, holding the oxygen near her nose again. Soon she is breathing easier, and the pink returns to her gums. Yet as the hours tick by, Bella's breathing grows more labored, and Lisa gently mentions that my plan to take Bella to the park may not be as ideal as I have envisioned. I can see Lisa's hesitation in telling me this. But it makes sense. Bella will need to be taken off oxygen, driven to the river, and set up again out there, and all this may cause more stress for her. I'm having a hard time knowing we're in a vet's office, where I'd vowed we wouldn't end things.

Jordon arrives just before midnight, and I feel so grateful. He stays with me for a few hours, so I have a close friend near me too. I thank him for coming out, and he shrugs off the gratitude: "That's what brothers are for." After a while, Jax decides to leave. He must consider his job finished. Jordon decides to leave too, so that Bella and I can have our last moments together. It's nearly dawn.

I am so full of emotion, I need to write. I pull out my phone and use a talk-to-text function so that I can still give Bella my full attention.

I am lying next to Bella in the "comfort room" of Four Seasons veterinary clinic. The time is nearing to assist her on her journey to the other side. She's calm and peaceful right now, and after a night of labored breathing, I want to cherish this peace and calm.

I've often said that I've been waiting for what this journey is teaching me. Perhaps it is this moment right at the very end that will teach me the most.

I want to get her to the park that's thirty minutes away and allow her one last chance to run around off leash and get into the water — to let her be a dog as much as she can. Let her spirit be free. Let her nose sniff as she explores. Yet, if taking her off of oxygen for the thirty minutes only causes her more stress, is it worth it?

I lie next to her and pay full attention to her body, her spirit, her energy. I pet her silky-smooth fur, starting at the top of her skull just above her eyes, taking a moment to let her ears run between my index finger and my thumb, letting them drop, playing with that little pocket in the ear on the anterior side, near where it connects to her head. I've always loved playing with that little spot and wondering what purpose it serves.

I run my hand down her neck and feel her solitary broad shoulder blade; the shoulder that has done so much work over the last eighteen months. I bring my hand down and massage the remarkably profound muscles that are developed in that lone front leg. That beautiful front paw that has borne so much weight and touched the ground in so many places in this country. Though she has a decent amount of gray on her face, it's nothing like how gray her front paw has become.

Almost every toe is completely white with a small band of chocolate that runs across the middle of the knuckles. The little hairs that stick out between the pads have been white for what seems like as long as I can remember. Every so often I still tickle them, just to be a pest. Funny how we show our love sometimes, isn't it?

Bella moves her head up and looks at me. She has a hard time getting comfortable again as she tries to lay her head back down on a pillow. I take her outside to go potty once more. We're only gone for a short time, but again her gums change color from the lack of oxygen. She lies down and pain meds are hooked back up to the port taped around her front leg. As I hold the oxygen to her snout with the one hand, the other hand returns to petting that ear. There's just something about the way it glides through my fingers. I let out a deep sigh. I sure am going to miss that feeling.

I've moved back to her shoulder blade now, tracing her smooth coat down her spine and to her hips. I massage down her rear left leg and trace my fingers around the fatty lipoma that's grown to the size of a Ping-Pong ball on her knee. I can't count the number of times people have asked me about that growth. My Ranger buddy from Savannah so aptly named it her "lump of wisdom," which he would tickle as he loved on her.

Bella's curled in one of those semicircles that dogs often sleep in. I'm now pinching the skin between the tendon that runs down the back of her leg and connects to whatever you call the anatomy of this backward knee on the hind legs. It always fascinates me how seemingly vulnerable this tendon is. As I trace my hand up to the pads of her left foot, I find my hand next to her face again. My thumb accidentally moves under her ear and tickles enough to make it twitch. I'm reminded of the times I'd blow in her ear, knowing she didn't like it. But we were like kids picking on each other on the playground. Every tease means *I like you,* right?

My hand is now up near her jowls, and I wish I could pull on them for fun like I used to. But after the osteosarcoma attacked her mouth, pulling on her jowls wouldn't be pestering, it would be torture. I think again of that shapeless brown birthmark, the one that's our little secret, hidden underneath her left jowl. I don't have to see it to know it's there.

I admire the gray around her muzzle. Time has treated her well. She's nearly ten years old, but I wouldn't put her in a day over seven. Wait . . . she's listening . . . make that five.

Her perfect nose. The dry Colorado climate and perhaps some of her treatments have dried the top out a little bit. I've been trying different remedies — coconut oil, a tiny bit of Bag Balm — to clear up the chapped appearance. It's been disappointing not to be able to kiss that nose lately; I've had to relocate my landing point to her forehead. She's had no lack of kisses.

I like running my middle finger from the middle of her snout up the valley between her eyes and over the ridge of her skull. What a hard noggin she has. Lord knows I've seen her bonk it into plenty of things, and she's never made a peep. A memory of her nearly knocking me out with that thick skull of hers causes me to laugh out loud. Bella's eyes slowly open and close.

Those copper-colored eyes.

Bella breathes in. Breathes out. Breathes in.

She is still with me.

An hour ticks by, and Bella is resting peacefully. I find I'm not as anxious as I once was. Is this my biggest lesson from Bella? To not get so wrapped up in the story and fret so much about how life might unfold, but to simply embrace it and be grateful for what it is. To be more at peace.

I continue petting her near that area on the rump that every dog owner knows is the magic spot. I give it a gentle scratch to see if I get a response. Nothing. I want to scratch harder, but the joyous flail that it usually induces wouldn't be beneficial to this peaceful slumber that she's in right now.

Gently, I run the sides of my fingers along

her tail. I'm going to miss her tail the most. Her wagging swishing thumping flinging tail, this indicator of happiness. I laugh quietly, reminiscing about how straight it sticks up when she's chasing a squirrel up a tree, or when she'd help carry trash to the Dumpster at the apartment building in L.A. How the hair stands out at the base of it, and how it broadens along with the hackles of her neck and shoulders. Such a proud girl.

Someone shared with me a story about why it hurts so much when you lose your dog. The pain you feel is because of their wagging tail in your heart. I think of how hard Bella's tail swings back and forth. Damn. This is going to hurt. I've prepared myself for some pretty good pain in my heart, but I think it might just break.

I put my arm around Bella and drift into a quick slumber.

I'm transported out of the room. I'm dreaming. I see the outdoor patio of my favorite restaurant in Lincoln. I'm sitting at a table, with someone across from me. Our table has an umbrella, and the sun is setting at such an angle that its rays shine down on us. We're drinking something. Iced tea, I think. The remnants of two McQuinton sandwiches and tortilla chips with chili con queso adorn the plates. The person across from me is a man. Small to medium frame, about five feet seven inches tall. He's wearing a sweat-stained baseball cap, a well-worn long-sleeved T-shirt, khaki cargo pants, and hiking shoes. Under his hat are eyes as blue as the Nebraska sky.

It's Mike.

God, it's so good to see his face. I can't quite make out the words between us, but I can recognize the tone of his voice, the nature of the conversation. It's very casual, very matter-of-fact. I've dreamed of Mike before, but this time is different. He is on the other side, and I am just visiting. We are both aware of this, and we both know why we're having this conversation. I have come to the table today to tell Mike that Bella is crossing over. Mike is here to reassure me he'll watch her until I get there too. I hand Mike Bella's pink leash. It's a changing of the guard. A transfer of guardianship. "Don't worry," he says — and these words are clear. "She won't need that here."

Before I can get up to hug and thank him, a gentle knock sounds on the door. I open my eyes and glance at the time on my phone. I've been sleeping for only about thirty minutes. Lisa comes into the room. She has awakened me from the dream.

Lisa gives Bella a dose of pain medicine and nods to me. We won't be disturbed again for a few hours. I lie back down again with Bella, thinking of where she'll soon be going and who'll be there waiting for her when she arrives.

16
From Beginning to End

Dawn is breaking. I know now that Bella will breathe her last here at the vet's. She won't be at her favorite park. But maybe, at the very least, she can be outside.

The team has told me they will be back to help Bella pass at seven thirty A.M., at sunrise. The hour is drawing closer. Bella is still asleep on her bed in front of me. She breathes in. Breathes out.

There's another gentle knock on the door. Kathryn, another friend, who works the front desk at the clinic, has arrived. I tell her of my decision to take Bella outside to die, and Kathryn goes off to set up blankets in a grassy area of the clinic's yard, the place we both think will be best. Lisa returns to the room one last time, not to check on us, but to tell us that everything is ready.

I stand up and ask Bella one final time if she wants to go outside. She rises, a little wobbly on her three legs, and follows me down the hallway and through the doorway to the plush grass in between the buildings. Kathryn has laid out a large blanket for Bella and me to lie on.

Bella struggles to breathe. She walks around, almost wandering, and finds a place to go potty. She perks up, sniffs the breeze. All night she's looked so close to death, but outside she seems to gain a second wind. "I've heard this happens a lot," I say, to no one. Animals often have a little surge of life at the very end that makes the decision to put them down all the more difficult. I think: *Oh, Bella . . . I'm sure you could run and explore forever. But you're suffocating by the second. Precisely the reason they don't want me to even transport you thirty minutes from here. You could die a painful death in the back of the 4Runner while I'm driving . . . I think this is best for you, baby. Please know I'm doing my best for you.*

Gently, I call her over to the blanket. Bella lies down, and I lie next to her. The vet who'll be doing the procedure approaches us. He's come into the clinic during the morning shift change. He's a larger man, early forties, and Bella and I have not met him before. But he is friendly and kind and I can tell it's going to be hard for him to do what he's come to do, particularly after just meeting us.

Kathryn gives me a white paper bag with a bacon cheeseburger inside. One final treat. I give Bella a little bite and then put the burger in my mouth and hold it close to Bella, teasing her one last time. She leans toward me, grabs the burger out of my mouth, and eats it in one fell swoop. She's never had bacon, or mayonnaise or cheese, and sadly I don't think she even tastes this bite. She gobbles it down and is ready to eat a dozen more. If only she could breathe.

The grassy area slowly grows brighter as the sun works its way above the horizon. The vet prepares the solution, and Bella tries to stand. Lisa and I tell her to lie back down, and this crushes me. Bella wants to get up. She wants to have a normal fun day. I pet Bella's head and reassure her that everything is going to be okay. Bella sits in a sort of sphinx pose with her feet under her in such a way that she can remain alert and rise if needed. She looks around with a rapid and heavy pant. Her eyes look confused, and I try to calm her. Her confusion isn't because we're making her lie down; the confusion is due to the lack of oxygen to her brain. We've only been outside for a few minutes, but those few minutes away from oxygen are taking their toll. This is what Lisa is trying to save Bella from, to save me from. I'm suddenly glad we didn't try to go to the park. I'm suddenly reassured. I don't want to watch her suffocate to death.

The vet holds Bella's right leg and presents a syringe filled with white, chalky medicine. Propofol. It's often used as a sedative for people, and I've seen it before during my EMT clinical hours in the emergency room. Although I know how quickly it can take effect, I am not prepared for what's coming next.

I lie down by Bella's side, and when the vet asks if I am ready, I say yes. Then I decide to hop up in front of Bella so I can face her directly. So we can look into each other's eyes as she passes.

"Rob, this is going to happen quickly," Lisa says. Her tone is sharp. "You're going to need to catch her!"

My heart races and I quickly hop back to Bella's right side. I place my hands on the sides of her face to look into her eyes. She looks just past my direct eyeline, still panting heavily. Then, as the solution goes in, her eyes dilate in an instant, the neck muscles that hold up her head relax, and her head and body fall into me.

I quickly shove my right arm under her

belly, my left arm over the top of her chest. Her missing leg always made for such convenient cuddling. I hold her as tight as I can, and I run my hands up and down her body, knowing this is the last time I'll ever get to pet her.

The vet injects the second solution, the one designed to stop the heart.

I hold Bella tighter, attempting to pour every bit of love and energy I have back into her.

The vet listens with his stethoscope. He pulls it away, and I ask with a cracking voice, "Is she gone?"

He looks at me solemnly. He nods his head.

"Oh God," I say. "My baby. Oh my God, my baby. I'm so sorry. I'm so sorry, baby. Thank you. Thank you for everything." I start bawling uncontrollably. I didn't think I would. I thought I'd be strong enough to witness it happen and breathe deeply, knowing this is all part of the process, knowing that she is at peace. But this all happened so fast. She seemed so alive not even a day ago. So alert, just hours ago. Still breathing, just seconds ago.

Bella is gone.

My body shakes as the emotions that have been built for more than a decade explode from my spirit. My life will be changed forever. Bella is lifeless. Gone from my physical world. No more wet nose in the morning. No more wagging tail. No more barking every day at five P.M. to remind me it's time for dinner. No more taking my keys and asking for a ride. No more chasing squirrels. No more face licks. No more hours spent wasting the day away cuddling on the couch watching Netflix. And the strongest connection I've ever had with another being is gone.

The pain is so heavy. It hurts so badly. Unlike anything I've ever imagined. The deaths of my brother and sister weighed on me so much, but the immediate stab of Bella's passing feels so unbearably sharp. And I had to make the decision to do it.

My bawling is heavy and ferocious. Nothing's held back. Tears keep flowing from my eyes, snot runs out of my nose. I'm not sure how much time passes. I bawl until my eyes are dry. My breathing settles. My mind clears. Bella is gone. Finally, I look up and around at the staff, and everyone outside is bawling along with me, including the vet who's met us only this morning.

It's an incredible moment, yet I feel cheated. *This can't be it.* I didn't actually *feel* anything the moment she passed. I didn't sense her spirit leaving her body. There was no vision of a ghostly version of her running into the horizon. There was no ribbon of energy, visible or palpable, that emitted from her body and disappeared into the ether. I've experienced the loss of loved ones, but this is the first loss I've experienced first-hand, and suddenly I'm questioning if there is anything after this physical life. I've always believed there is something next. I need there to be. But given how strong my connection with Bella is, if I can't feel or sense that transfer with her death then I can't help but lose faith that anything happens at all. Maybe we just die, and that's it.

Then, as I'm standing above her lifeless body, wondering what happened to that amazing spirit and reeling over the fact that her life just disappeared, a gust of wind rustles the leaves on the ground. The wind picks up and suddenly a small funnel of gold and yellow leaves is lifted into the air. This little leaf tornado circles Bella's body, and the leaves fall at my feet as I feel the wind blow gently across my face.

Ahh. There you are, baby girl.

Lisa asks if I want help carrying Bella's body inside to the X-ray table so we can get paw prints made. We had one made when Bella first came to the clinic, but the print itself is too shallow. You can get a better print after a dog is gone, because you can press harder with the paw.

"No, I'd like to carry her myself," I say.

I pick her up, and her lifeless body feels so different. So much heavier. I know that's why Lisa asked to help. But I want to feel Bella's weight myself.

I carry her inside and lay her on the table. We make four paw prints out of clay kits. I want two for myself in case one breaks, one for my mom's house, and one for Charli. It takes three clay kits to make it all happen.

Lisa asks if I want them to take Bella to the crematorium.

"No, I'd like to do that, please," I reply.

I'm not leaving her side until I absolutely have to. Again, I choose to carry her outside by myself, to feel that weight again, her body still warm.

Lisa opens the rear door of the 4Runner and crawls inside to help me get Bella settled. We place blue-and-white pee pads on top of her bed to absorb any fluids that might leak out. Once Bella is inside, I turn and hug Lisa, her face still wet and red from bawling alongside me. There's something so comforting knowing that someone else is sharing the pain my heart is feeling, to know I'm not alone in this.

"Thank you," I say simply, and hug her again. I shut the rear doors and get behind the wheel.

Kathryn knows the way to the crematorium and offers to escort us. She leads the way in her car as we trail close behind. Thirty minutes northeast is a rural crematorium called Precious Memories. There's something majestic about the location, a few acres of sacred ground where so many people have said goodbye to their loved ones. Outside the main building is a cemetery for those who wish to bury their animals and have a headstone.

When we get there, we park, and I leave Bella in Ruthie. Kathryn and I walk into the office and meet the man at the front desk, who's already expecting us. As he prints paperwork to be signed and shows me the choices for urns, his tone is somber and respectful of my loss. There are beautiful wooden urns that would be a perfect fit for Bella — she loved nature, especially sticks to chew on — yet they'd be permanently sealed. I want an urn that can easily be opened so I can periodically scatter her ashes throughout my continued travels. I will take her to the places I'd planned to go and leave a little of her remains along the way. I promised I'd get her to the Pacific Northwest, and I vow to fulfill that promise. Finish what I started. A copper urn with two paw prints on the lid shines like the holy grail. The color of the urn reminds me of Bella's eyes — those adoring, beautiful copper eyes. It's the one for us. I also purchase a small key-chain vessel so I can always have a part of Bella with me wherever I go.

The man tells me I can pull into the bay and unload Bella onto a cart, but I say I would rather carry her. I want her in my arms. I want to feel her weight one last time. I open the door to Ruthie and gather Bella up. Her tongue is turning blue, but her body is still warm and soft.

"I got you, girl," I say. "I got you."

I had always assured her that I had her back when she needed it. She learned to look at me with a certain expression that said: "Hey, I can't do this one. . . . A little help here?" She didn't need help often, but I always felt more than happy to swoop in and be her hero. Today, at the crematorium, I will be her hero one last time. In the bay is a large plastic bin with a dog blanket. It's patterned, black, red, and white, and has bones and paw prints, and the word *woof* printed. I could bring the cart to the 4Runner, but I owe her more than that. I don't want to see another loved one on a cart. I carry Bella the twenty yards from Ruthie to the bay and lay her down in the bin.

This is it. My final goodbye to the body that housed her spirit.

The man comes out of the office and asks if I'm ready for him to take her, or if I need more time. I want to be strong enough to say I'm ready. But I am not ready. "I'd like more time," I say, my voice cracking. Something isn't right. I can't just leave her here like this. I know I need to do something else to mark the moment, but I don't know what else to do.

I run back to the 4Runner and get out her favorite play rope, one with giant knots in the end. She's had it for years. It's the kind of toy you have to warn people about, because it can do some serious damage to knickknacks and glass doors in entertainment centers. I also grab a chewed-up Chuckit! Frisbee with a hollow center like a doughnut. Whenever she caught it, her snout went through the hole, and the top would cover her eyes as she ran around blindly after retrieving it. I had thought these toys might be great keepsakes for me to remember her by, yet the more I think about it, it'll be more fitting to send them off with her. Deep down, I have an idea that they'll make the next place feel a bit more like home.

I run back, show the man the toys, and ask him if these can be cremated with her or if they will only be tossed to the side. If that's the case, then I'd rather keep them. He assures me they will go with her, and again I'm glad we have such a great place for her to be. He also reassures me that I'll be getting the ashes of my animal, and only my animal. Some places do mass cremations and divvy up the remains to the owners. They charge you extra to get your animal's specific remains. Even then, I'd probably shell out the extra cash so I'd know I wouldn't be carrying Rex's ashes on my key chain, or spreading Fluffy's ashes in the ocean. What if Fluffy hated the water? Yeah, I'm glad to know it'll be all Bella.

I place Bella's favorite toys alongside her body. I say one final goodbye and turn to leave, but my heart is pounding, my breath rapid, my mind fluttering. I'm still not ready. I still need something more. I'm having some kind of panic attack, and I recall one of the most important tools I've been given from my time with Warriors' Ascent . . . my breath. I can't leave while I'm like this. I need to leave feeling clear and in control. I don't want my last moments with Bella to be panicked. I need to have a ceremony with her to gain that closure. The toys to take with her are a good start, but I need more.

I turn back toward Bella, take a knee, place my right hand on her chest, and become aware and conscious of my breath. As I examine her body one final time, I take several deep breaths. I pet her entire body as I continue to breathe deeply. I admire her perfection one last time, taking in all my favorite parts. The spiral of her fur that grew in place of her missing leg is one of my favorite physical characteristics. That missing leg was the source of so much of the inspiration she's provided over the last eighteen months.

I realize that something needs to be said. That's the missing piece to the ceremony. Still down on one knee, my hand finds its way to Bella's ribs. I close my eyes and continue with my even breathing as I say: "I send all my love with you, forever and always. I ask that you send yours back to me, so that I may carry it with me, forever and always."

My anxiety lifts like a cloud.

Just like that.

My lungs are able to expand fully, and I can now breathe and see clearly. I kiss Bella's forehead and then remember that her beautiful nose, which had been too painful for me to kiss, is now pain free. I lean over, give her one last long kiss, and whisper, "I love you, baby. Thank you for everything."

A worker appears in the bay, and I say that I'm ready. We speak for a moment about Bella's amazing adventures, and I give her a hug and walk to my car.

I have made my peace. I am calm. I pull out of the driveway knowing I have done the one thing that I promised to do: I stayed with Bella until the end. I followed through. I finished what I started all those years ago. I gave her the gift of being there while she passed, and she gave me the gift of experiencing a life from beginning to end.

My phone rings.

It's Maria, the Argentinian woman who'd welcomed me and Bella into her and her husband's home. I tell her that Bella is gone, and in an instant Maria's bawling. Once I hear the pain in her voice, I join right in. She asks if I want to come to their place, and I do — I don't want to be alone. When I get to the apartment complex, I start to text Maria: "We're here . . ." Then I delete and start over: "*I'm* here." Bella and I were such a team. Rob and Bella. I even signed my emails that way. It hits me like a dull knife in my gut. There's no more "we."

Maria hugs me with all her might. Her dogs, Miyu and Pepe, both rescues from Argentina, notice my energy right away and lean their heads into me, giving me all the love they know how to give. It feels both beautiful and heartbreaking at the same time.

Our friend Tad comes over with his loving rescue shepherd mix, Luna, and she follows suit with the comforting, but with a bit of higher puppy energy, since she's barely a year old herself. When Maria's husband, Jon, comes home, he gives me a hug too, and we decide to get out of the house and grab something to eat. We choose a Thai restaurant within walking distance, and I eat, and the food tastes good to me, which surprises me. I've been thinking nothing will ever be right again.

We head back to their place and watch a movie, and when the show's over, I fall asleep quickly on their couch. The next morning when I awake, all the peace that I've gained during the ceremony of saying goodbye to Bella has left me. The images of her confused eyes dilating and her body falling into me haunt me. I replay these images in my mind, and I can still see them so clearly. Over and over. I feel I failed her. I could've taken her to the water. I feel I was rushed at the vet's. Suddenly, I forget all their graciousness and kindness. I feel resentment. Anger. Blame. Surely they pressured me into it. My breath grows rapid and my heart races again. *What have I done?*

Maria is up, and I tell her what I'm feeling, and she all but yells at me: "Rob, oh my God! You did everything for her! Did you want her to suffocate to death?"

"I would've rather she ran until she couldn't run any longer!" I say — and all my thoughts tumble out in a rush — "I wanted her to play in the water one last time. I should've let her decide! I shouldn't have gone to Denver! I should've stayed here in Fort Collins! I shouldn't have woken her up to take her outside! I should've done it in the room! She was confused! We made her lie down. We *made* her lie down!" I am so sad, so angry, so *guilty.*

Maria shakes her head. She's emphatic. "Rob, you did everything for her."

But I'm crying again. "I could've done more."

■ ■ ■ ■

In the afternoon, one of the vet students I've met through Lisa texts me to ask if I want to watch a movie that night and, in the morning, go to Rocky Mountain National Park. I text her back immediately that it's a good idea.

I drive over to her place, and she's got a rescue blue nose pitty named Poppa, who's battling terminal cancer. My friend Mariel refers to her dog as her "guardian angel" and tells me that Poppa has always slept in the same bed with her. Poppa snuggles up close to me on the futon, and we all watch a fun movie together, and when my friend turns in for bed, Poppa starts to follow her, but my friend says, "No, no. You stay out here with Rob." I call Poppa back up on the futon, and he hops right up and curls up next to me. I look at my friend and thank her. I know how costly this gift is for her. How I feel about Bella is how she feels about Poppa. She knows the pain of seeing a dog slip away and how much love a dog can give to a person who's grieving. She gets it, like so many dog owners do. Having that common understanding is a bond all its own.

In the morning, on our way toward Estes Park, we stop by the crematorium and pick up Bella's remains. It feels so great to have part of her back. Though her spirit has crossed over, there's something truly special about having what's left of her in my possession, a tangible memory to hold in my hands.

My friend and I drive to the national park, Bear Lake. Signs say parking is full, so we pull into the public lot below and take the shuttle bus to the lake. A couple with three young children approach us in the bus, and one of the little girls, maybe six years old, asks if she can sit on the seat next to me. I look at Mom for the go-ahead, and she gives a welcoming nod. The little girl hops up right next to me and introduces herself.

"What's in there?" she asks, pointing at Bella's urn.

Again, I look to her mom and again am given a nod of approval.

"Well, this is my dog, Bella," I say. "She's gone now, but her ashes are in here."

"How'd she get in there?" the girl asks. "What are ashes?"

I look at the mother a third time, hoping for a little guidance on how she wants her daughter to learn about what happens at the end of life. The mom says simply, "Go ahead and explain it to her," so I try my very best to do so without traumatizing the poor girl. She's not fazed at all. Only curious. It feels wonderful to be in the presence of such innocence. My friend and I talk to the family and learn the mother's from Scotland and the father's from Ireland. We talk about work, life, and of course dogs.

When the bus arrives at the lake, we give the family hugs and pose for a quick picture before parting company to enjoy the park on our own. I have things I need to do. The lake's only a short walk away, and I quickly find a rock to sit on and place Bella's ashes next to me. We're near the water, and I find my breath again and glance from the copper urn to the lake. Beautiful mountain peaks serve as a majestic backdrop, and I can feel Bella in all the essence of the wilderness. I ask Bella for forgiveness for being on my phone too much over the past few weeks, for taking her to Denver instead of snuggling with her in her favorite leaf pile, for making the decision to end her life if she wasn't ready. I tell her I'm sorry if she left this world confused. I'm feeling guilt, whether warranted or not.

Again, I need some sort of ceremony now that she's here physically. And it's almost like Bella is speaking to me. Sitting here on the rocky lakeshore, I feel a calmness. A peace. Bella is in the peace, and I feel her energy telling me, "I may have been confused in those last moments, Daddy. I wasn't sure what was happening. I was struggling just to breathe, I was in pain. But I'm not struggling now. I'm not in pain, and I'm not confused. I am at peace, and it's so beautiful. I know you did everything you could. I know you loved me with all your heart. All I want is peace for you too now, Daddy. Feel this peace? It's so beautiful, isn't it?"

Is she communicating those exact words to me from a world beyond? Most likely not. Yet I believe — lots of people believe — that once a life has crossed into the next phase of existence, our spirits can communicate in a way far deeper than earthly limitations allow. So I pay attention to the energy I feel and put words to it in a way that will help me heal.

Before we leave, I take a few photos of Bella's urn with me, the way that we took photos together when she was alive. It just feels like something I need to do. I step out onto a rock, open her urn, take out a solitary pinch of her ashes, and sprinkle them into the clear water. I watch the small particles dissolve. Little bone flakes drift slowly to the lake bed. I have made her an eternal part of the land. She would have liked that.

We ride the bus down to the parking lot and trek back toward Fort Collins.

I drive to Denver and spend three weeks with friends. We go to football games, and celebrate the Marine Corps' birthday, and go mountain biking. Veterans Day passes, and I go on a few dates, and I meet with vets getting hyperbaric treatment for PTSD and traumatic brain injury to see if I can help.

It's a great few weeks of welcome distraction. But I feel I have some unfinished business yet to do. I don't want to wallow in the agony of Bella's loss forever. Yet I do want to complete the process of my tribute to her life.

I need to finish what we started.

I need to do it for Bella.

I need to do it for me.

17
The End of the Line

I'm still a little lost.

Friends in the Denver area invite me to stay in the spare bedroom at their house. I place Bella's urn on top of the dresser in the bedroom. Evening falls, and I am tired, but before I climb into bed, I hug Bella's urn. I tell her goodnight. I kiss the paw print on the lid. "Good night baby girl." Near the nightstand is Bella's indoor humidifier, and I turn it on to hear the hum, see the familiar steam rise. The little blue LED light makes the water glow, and I turn off the room lights and lie on the bed. I'm on my side with my head on the pillow, and I stare at the LED light as if it's a night-light. The steam dances its way into the air. Just like Bella's spirit. It's as though she is here in this room with me.

Days pass, and I know I need to complete this mission, need to reach Oregon, but I decide to wait to return to the road until after the New Year. I want to head back to Nebraska first and spend the holidays with my family. But I need to be careful, too. Because I don't want to go back to Nebraska and settle there. Not yet, anyway. I want to keep momentum and keep this journey going until we finish it.

Even then, it takes me a while to get going. For days, I stay around Denver. I'd like to think our journey helped with my depression, but now I'm not sure. The depression hasn't disappeared but has been muted for a while now, pushed aside by the sense of purpose our journey had given me, that Bella had given me. Now that Bella is gone, that purpose is fading, and the depression works its way back toward the surface. I want to do things for the right reasons. I want to keep moving forward, keep living for the purpose of helping others, a purpose that a mere few weeks ago seemed so clear. I want to live in the moment, gain perspective, and inspire others to live their lives well. I don't want to get lost again. But I'm struggling to be in the moment because I'm worried about the future and depressed about the past.

Breathe, Robert . . . breathe.

I guess I feel guilty. Is that it? There's also a sense of relief about not having to worry if Bella can come where I'm going. Not having to worry about fitting my plans around her. Not having to worry about her health. But then I feel guilty for feeling relieved. I talk to people about this guilt, and they say it's completely normal. They too were once in a caregiver role with a dog, or maybe with an elderly relative. Then the dog passed, or the grandmother died, and they felt bereft yet guilty because they also felt released.

I try to hear Bella's voice, still in my ears. At night, I keep turning on the humidifier.

"Daddy, keep living," Bella whispers through the steam. Or maybe this is my own inner voice this time. *You lost a loved one, and you're mourning a life that's now completed. But you didn't lose your life. Your life is still incomplete. Keep going. Bella loved being alive: don't you think she would want you to keep living too?*

End of November, Thanksgiving, and I am at home with my mom. We eat dinner with one side of the family, then head to my nephew Andrew's place and have dessert and play a card game called Cards Against Humanity. We each toss back a couple of Moscow Mules, a delicious ginger beer–and-vodka cocktail. Andrew's girlfriend is here, and Chandler is here, and the game sets up everybody to say the most indiscreet things. Everybody's laughing, and Mom plays too but says the words under her breath.

"I've never said these words in my life," she says.

"Sure, sure, Mom," I say. And we all howl like hyenas.

The game finishes, and the playful energy has made us completely relaxed with each other. There's a lull in the room, then Andrew becomes serious and says: "Uncle Bob, ever since Mike died, I've seen you doing double duty as our uncle. I just want to say I notice it, and I appreciate it."

Tears well up in my eyes. Mike was such a damned good uncle, and I always felt we worked as a team to teach our nephews the life lessons that uncles are allowed to teach — the kind of stuff that parents can't get away with. Since losing Mike, I've done my best to make up for what my nephews could no longer have when they lost him. I know I can't fully make up for his absence, but I've tried. Honestly, until this moment I've felt like I've done a lackluster job, being gone so often, but Andrew's words fill one of the many holes in my heart.

Mom clears her throat. "Bob, what I've noticed is that you've stepped up and taken

Mike's role as the caretaker of the family."

This hits me like a freight train. I don't want to bawl like a baby, so I straighten my shoulders, my jaw set. A few tears roll down my cheek, and I say, "I'm trying." I know I do a lot of things for myself, trying to make sense of my life, but I really do try to bring those lessons back home. With so much of what I do, even when it seems to be just about me, I have my loved ones in mind. And caring for Bella has taught me a thing or two about being reliable and kind, generous and selfless.

A dam breaks in our family. We all sit back and share our memories of Mike. We share the pain of our loss, yet we relive the happiness, too. It's been a ten-year buildup of emotions, and it feels healing to release these memories at last.

Mom cries too. She cries freely. It's been nearly a decade since Mike's death, when she told me she hadn't been able to cry. I had assured her that someday she would, and as much as it hurts to see your own mother cry, I must tell you that it was a beautiful sight to see her let out her pain and heartache that had been harboring inside for years. Today she cries the tears of grief for Mike, the son she lost. Perhaps, too, she is crying tears of happiness for Rob, the son who's found. Old wounds have finally healed over. Mended. I'm no longer the lost child, always stretched between two families. I have a role and my family sees it as valuable. I'm needed. I'm a caregiver. Mom has told me that I am now the family glue that Mike once was. And I can't help but think I've got Bella to thank for this turn of events.

For a long time, I thought of Bella as my child. Over time, she became my friend, my companion, my copilot, and my soul mate. But it took me a long time to realize she was also my mentor. She taught me all about unconditional love. She taught me about forgiveness and loyalty, and not to judge. Whatever passed between us that was less than perfect, she knew how to move on to the next moment. We could look into each other's eyes and know that we were there for each other, through all the ups and downs. We always had each other's backs.

I loved that feeling. And the best part about it was that it taught me to do the things I find most difficult — be myself, forgive myself, and love myself.

In my old bedroom in the basement of my mom's house sits a built-in cabinet, and I ask Mom if I can remove the blankets in there so I can turn the cabinet into a tribute space from my travels. She says sure, so I place my military shadow box on the top, my college diploma, and my NCO Marine sword. I fill in the gaps with treasures I've collected along our adventure. A couple of smooth gray rocks from Put-in-Bay. Pinecones from the Adirondacks. A stone from Acadia National Park. Seashells from the Carolinas and the Keys.

In the center sits the beautiful copper urn with the solitary paw print on top. I keep Bella's urn with most of the ashes on the cabinet, then fill a smaller container with about a quarter of what's in the urn. I'll use this to spread her ashes on the last leg of our journey together. Bella would approve of this, for sure.

We are set. We can finish our trip together.

I set out again in early February 2017. Ruthie feels strangely empty without Bella beside me. But I have the container of her ashes, and my heart is full.

We stop in Lee Martinez Park in Colorado, where Bella found her favorite spot to lie in the water so many times in the last few days of her life, and I let a pinch of Bella's ashes go into the river. I drive up to Horsetooth and Dillon reservoirs, the big lakes where Bella and I spent so many happy hours on the paddleboard. I let a pinch of her ashes go in each.

I drive straight through to Arizona, to the Grand Canyon. It's freezing, reminiscent of last year's winter travels in the northeastern states, and at one A.M. I pull into a campsite that's marked CLOSED and sleep for a few hours before sunrise. My alarm buzzes at six A.M. Bella is not with me to lap my face, so I hit snooze once, twice, then remember why I'm here. I drive out of the campsite, looking for a viewpoint.

At a desert watchtower that's constructed out of stone, I stop. In the distance, I glimpse the vast canyon, but it's not until I walk closer to the south rim that I'm truly able to appreciate the colossal scale of this natural masterpiece. It's desolate this time of year, and in this moment, the canyon belongs to Bella and me. Two ravens watch curiously as I gather snow and pack it with a pinch of powdery ashes. I stare at the snowball a moment, then just as the sun rises above the horizon, hurl it deep into the canyon as far as my arm can launch it.

We drive on to Death Valley, on the border of California and Nevada. Badwater Basin is my first stop, the lowest point in North America. I park Ruthie and walk down a wooden stairway that leads toward the salt basin. Badwater Basin is bleak, forsaken, like a scene from a movie where someone's crawling on all fours, skin blistered, searching for a drink of water. I walk out into the flats about a hundred yards and pick up some salt from the ground. I put it to my tongue, just for the experience. It's the strongest salt I've ever tasted. I sprinkle some of Bella's ashes onto the dry land. She blends in with the chalky desert floor, and it feels symbolic. The basin is literally rock bottom, yet the best thing about being at rock bottom is that the only way is . . . up.

We travel to California, so green right now, thanks to what people are calling the twenty-year snow. A longstanding drought has officially ended, and the hills look more like Ireland than California. I make friends with strangers and together we witness the phenomenon called "firefall" in Yosemite National Park, where I place a pinch of ashes in a flowing stream. I visit a friend in Lake Tahoe. The same weather that brought the rains to central California brought record-breaking snow to the Sierra Nevada range. Her family has a little cabin right on the water, and my friend and I walk out onto the dock, and I ask her if I can place some of Bella's ashes in the water. She says she'd be honored, as this is the final resting place for the ashes of her own family's dogs. "Well then, looks like Bella will have some friends to play with here," I say as I sprinkle a few ashes and again watch them dance through the water and eventually become one with the vibrant blue lake.

Heading north toward Oregon, I stop by Shasta Dam and leave a pinch of Bella in the waters there too. The whole area looks like one gigantic swimming hole. Something Bella would have loved.

The border between California and Oregon is just up the freeway.

Our final goal is twenty minutes ahead. My heart pounds with anticipation, although I have nothing planned for when we arrive. All I know is I'm going to see the sign and drive across this imaginary border and it'll be mission accomplished.

I see the sign. WELCOME TO OREGON.

I pull over. I hop out of Ruthie and do a little dance, make my mark in the dirt. "Look, Bella!" I exclaim. "We made it! It took so long, but we made it, you and me!"

I consider spreading a bit of Bella's ashes right then and there, but although this moment feels magical, it's still just a spot of dirt alongside a busy freeway near an imaginary line. I decide to wait until I make it to the iconic Oregon coastline that I've always dreamed of taking her to. I'll save the moment for the ocean. Still, I'm happy. Ecstatic. Our trip is all but over.

18
Come, Find Me on the Beach

My plan is to head to Portland to visit friends, then head west to the coast, so Bella's ashes will be in the Pacific Ocean, too. But in Portland, I receive a sudden, warm message from a girl on Instagram. She doesn't have many photos of herself online, but from what I can see she is maybe late twenties, early thirties, and she's taken her own trip around the states with her golden retriever named Franklin Waffles, who's recently passed away. The girl's travel pictures are posted online, and they are incredible. Her love for Franklin parallels mine for Bella, if not even greater. Videos on her blog show her singing in the car to classic rock while Franklin sleeps mostly in the back, but with his head on the console or her lap. We've corresponded a few times already, as we were both traveling with our dogs at the same time, but never close enough to meet up. Part of my newfound purpose lately has been talking with people about pets they've lost; it's therapeutic. The conversations start with pets, then usually evolve into long talks about life, adventure, and dreams of all kinds.

"I see you're in Portland," she writes. "I'm in Hood River, if you'd like to grab dinner."

Hood River is east, about ninety minutes from Portland. It's St. Patrick's Day, she's watching a basketball game at a bar in town, and that sounds like fun. So I set the GPS for the bar and head east on Interstate 84. In Hood River we meet, and she says hi with the sweetest tone and a giant dimple in her right cheek, and I think, *Oh no, she's cute. Stay on task, Rob. We're here to talk about her lost pup.*

We mesh right away, and we walk downtown in the rain without an umbrella, because apparently if you use an umbrella in Oregon it's as mockable as wearing a life jacket in the shallow end of a swimming pool. We stop where a band is playing under a big white tarp and listen for a while to the funky music, then go to grab a little grub. "You okay with pizza?" she asks. I like her already.

Over dinner we talk about our lives and our adventures on our respective trips. When I learn the heart of her story, I'm captivated. She once had a six-figure job for a big-city corporation, but her best friend was diagnosed with cancer. She cared for her friend until the very end, then suddenly life as corporate executive seemed insignificant, unfulfilling. Ultimately she quit, started working as a freelance graphic designer, and took her seven-month road trip with Franklin. On her trip she also asked herself many of life's questions. She questioned her purpose and sought a new one. The answer that she found was to go back to school to become a nurse. She already has her bachelor's in business, but after being there until the end with her best friend, she realized she could make a difference in the lives of others. Not only does she want to have a new career, she wants to gain the skills to volunteer as a nurse. She wants to give her time away, helping others wherever, whenever, and I think, *Wow, this woman is incredible.*

I know we have our dogs in common, but I had no idea our commonalities ran so much deeper. To have both lost someone so close to us. To have that loss tear away the mask of what we've been told should matter in life. To use that new perspective to make a change. And to make that change by adventuring across the country with your furry best friend.

This night I crash on her couch, and in the morning we drive up to Mount Hood. It's still snowy on the roads up to the mountain. We talk about her favorite spots where she took Franklin hiking and swimming, and I snap away with my Canon Rebel T2i. It's the exact same camera she used to capture incredible pictures on her trip. The exact same. We drive in her little Mazda CX-5, a black four-door SUV. A bed is still made in the back, the one for Franklin, just like the one I'd made for Bella. I spot a hammock, the same brand and color as mine. The floorboards are dirty with gravel and pine needles. Baby wipes sit in the console. Yeah, this all seems too familiar. For the next two days, we keep touring Hood River and the surrounding area, and I feel so comfortable in her presence. It's like I've always known her. There's no reason to go anywhere else. But I remind myself I'm still on a mission. I've got to get to the coast. I've got to finish this thing.

When the weekend draws to a close, we give each other a little kiss just to thank each other for an amazing weekend. We both say goodbye, and I think we're both wondering if this is the last time we'll see each other. She works and is in school full-time, and I have plans to keep traveling, keep meeting people and their dogs, keep taking pictures and sharing their stories. I've made it around the States; now it's time to travel the world.

"If you make it to Cannon Beach, say hi to Franklin for me," she says as I climb into my 4Runner. "That's where his ashes are spread."

I get back on the road again, heading to the Oregon coast, wrestling with my feelings. What I've just experienced is so wonderful. So new. *This* is what I'm longing for. A person who gets it. A person who has had a shift in perspective after the death of a loved one. A person who took a long trip on the open road with her dog to figure out what to do with that perspective. A person who longs to live for what really matters.

Yet here I am . . . driving away.

I want to see as much of the Oregon coast as possible, so Ruthie takes me south to Eugene, then west to Florence. The scenic woods on the drive to the coast do not disappoint. The coast feels like another country. Wild, rocky beaches. Cold, crashing ocean waves. Sea lions and seagulls. Huge cedar trees and big-leaf maples. Black oaks and redwoods.

In Newport, I pull into a random motel for the night. Rain falls and I'm tired. I buy a bottle of wine, bring my stuff inside, open maps on my phone, and search for points of interest on the way up the Oregon coast. Cannon Beach.

I watch videos on her website, A Golden Road, of her trip with Franklin and smile as I watch her sing along to Lionel Ritchie as she traveled across the country. The words of the Girl from Hood River come to me again: "Say hi to Franklin for me."

I take a deep breath and imagine this girl saying goodbye to the ashes of her best friend in the ocean waters. My heart hurts for her loss, yet my heart is warmed, too. Her love for her four-legged soul mate is so obvious. I send her a text: "I'm planning on stopping by Cannon Beach. Would you like to meet me there and introduce Bella to Franklin?"

The mission is accepted. The following morning, we make plans to meet on the sand, in front of the ocean, on Cannon Beach, in about three hours. On my way up to Cannon Beach, I get a text: "Pull over and hike the sand dune in Pacific City!" I'm up for the challenge and attack the dune. The dune wins. But I still make it to the top. As the view levels out, I see powerful ocean waves slamming into giant rocks. I stand in awe of the spectacular sight, trying to wrap my mind around the magnitude of the ocean's power. I text the girl again. She's already in Cannon Beach. I've taken too long. I'm guessing I haven't made a good impression. "So sorry! Where are you exactly?"

She sends me a dropped pin on the GPS along with these words —

"Come, find me on the beach."

Who is this girl? Can she possibly be real?

I jump back into Ruthie, head up the coast to the spot where the GPS leads me, and park right behind her SUV.

The trail is sandy, I hear ocean waves, and my heart pounds with anticipation. When I reach the point where the trail meets the beach, I see a pair of faded, dirty, well-worn women's running shoes, the shoes of an adventurer. I'm sure they belong to her, and the romantic in me wonders if perhaps she's left them there as a clue. I try to trace her footprints, but they're soon lost. The view to the left is blocked by sand dunes. I look to the right and see the magnificent sight of a turbulent ocean. Waves crash onto gigantic ocean rocks. Seagulls fly and squawk, creating a cacophony above us. Yet there's no sign of the Girl from Hood River, and the anticipation builds. I hike closer to the beach. Past the dunes, the beach presents itself. The iconic Haystack Rock lies in the distance.

Suddenly, there she is.

She's sitting on a giant piece of driftwood, looking far off into the ocean, to the waters where she said goodbye to her Franklin. She has no phone in hand, no book, no camera, no stick to draw in the sand. She's simply sitting, appreciating all the beauty in front of her.

A part of me hopes she won't see me. I just want to stand a while and watch her hair blow in the wind. But she turns and looks my direction, and I greet her gaze with a smile. She stands and walks toward me. I'm hurrying now. She's hurrying too. We meet with open arms and pull toward each other with magnetic force. Our lips meet and stay there. The longer we kiss, the more right it feels. The slight mist falling from the sky forms into larger raindrops. Finally I pull away, look into her hazel eyes, and ask: "Are we in *The Notebook*?!" She laughs and rests her head on my shoulder and we continue the hug. *How great this feels.* My broken heart has found its missing piece.

When the hug finally ends, we hold hands and wander down the beach. I look at her with a smile, again wondering if she's real. Yet here we are, walking hand in hand as we trek the ocean shores.

"Time to introduce Bella to Franklin?" I ask, and reveal the small container of Bella's ashes.

She nods. "Yeah, let's do that."

We spot a rocky shore about fifty yards ahead and walk to a tide pool where Bella can join the ocean. This will be her first time back in the Pacific Ocean after spending so much time splashing in her younger years in Southern California.

"Do you think Franklin will play nice with Bella?" I ask.

"He was incredibly gentle, a bit of a scaredy-cat," she says.

"Well then, Bella" — I open the plastic container and take a large pinch — "I suppose you'd better play nice with Franklin." I drop Bella's ashes into the clear waters of the tide pool.

She has her hand on my back, and I don't feel I'm alone while saying goodbye to Bella. We watch the gray powder blend into the shallow waters, and the ocean suddenly rises and floods the pool, then recedes back beyond the rocks, taking Bella with it into its vastness. It's as if the ocean is accepting our gift.

"Did you see that? Did you feel it?" I ask her.

"Yes, yes I did." She squeezes her hand tighter over mine.

With the Girl from Hood River, I feel *possibility.* The possibility to love again.

My mind races, but I do my best to slow it down and not let the mind speak louder than the heart. The mind gets lost in the future. The heart feels the moment. The future is unknown, and I don't want it to get in the way of this moment. I don't know what the future will hold for this girl and me, but at the very least, I know again that love has not ended. There's always love.

Days pass, and at last we say our goodbyes. The girl and me. But it's not a final goodbye. We both know that. But I have one different, final goodbye I need to make, and for now, at least, that goodbye will need to come only from me.

Ruthie's windshield wipers throw the rain from the glass as fast as they can as I pull the 4Runner into the trailhead. I've been exploring the rain forest for two days now. Today's destination is a one-mile hike up the beach to a rock formation named Hole-in-the-Wall on the rugged Olympic Coast of Washington State. Locals have told me to go at low tide. I'll see countless tide pools, hosts to colorful anemones. A populated world that feels all alone, the locals say.

The sky is gray and powerful, but the rainfall is slight when I begin the hike. I've now spread Bella's ashes in more than a dozen places on our continued journey, but I've felt for some time that the process needs to come to an end. I want that end to have meaning. Today will be the place.

Bella's death has given me a new clarity to see what's important. What this entire trip has led me to is a freedom from expectation. I'm not defined by a job or title or degree or city or place in life. I can be and do many things, and it's possible to keep defining and redefining who I am. This journey has been about finding purpose, finding what matters, living in each moment, understanding what it means to love. Bella led me on a healing journey along a long road home where the healing happens slowly, along the way, little by little, bit by bit. And sure, there's more healing to come. But at least, for today, I am on my way.

Slick rocks cover the trail. Up the bank lies a fallen tree across the water. I duck under branches, legs brushing against ferns, and feel my way across a washout. Soon I am back on the trail again, then out of the forest canopy and onto the rocky shore.

I hike north for a while on the beach until I see giant rocks that jut out from the horizon. Tide pools dot the landscape, and inside are amazing anemones and green water flowers. The rocks form a towering wall that marks the end of the beach. Sure enough, a hole, large enough to fit a VW bus, lies overhead and in the rock. To reach the archway I must find a way to climb up the rock bed. It's about shoulder height, and a few calculated footsteps and handholds later, I'm walking underneath the arch. On the rock floor of the hole are tide pools, and pink algae is scattered on the rock. Bella always looked so good in pink.

"Okay, baby girl," I say. "This is the spot."

I reach into my pocket, pull out the container that holds her remaining ashes, and remove the lid. I gaze one last time at the powdery remains and take a pinch, feeling the texture as I squeeze my fingers together. Letting the ashes go into the pool, I watch them blend into the water. I've done this before, of course, but this time feels different. This is the last of our grand adventure. We've driven thousands of miles, swam on both coasts, hiked in several national parks, walked in countless major cities, and now our journey together just seems . . . complete.

It's time to be finished.

I tip the jar and pour out the rest of her ashes, setting her free. I can't help but let out a deep breath. Bella has adventured with me long enough, and it's time for me to find out what's next for me, on my own. It's time for me to let her go.

As I walk away from the wall and climb down back onto the beach, I imagine Bella running toward the ocean, just as she did not so long ago. I picture her swimming through the waves and disappearing into the ocean, where she belongs.

"Goodbye, Beautiful," I say out loud. I'm crying now. "Thank you for being everything I wish the world could be. Thank you for being my copilot. Thank you for being my best friend. Thank you for loving me. Daddy loves you, baby girl."

I take another deep breath.

I walk farther down the beach. *It's over.* What will become of me now? My mind races through countless ideas, and I'm excited, but I can't help noticing a sense of emptiness lies in my heart too. How am I possibly going to replace all the love that just poured out into the ocean?

I take a deep breath.

Then, as the tears flow stronger, Bella is

back. I feel her presence so strongly that I can almost see her hopping alongside me on the beach on her three legs. Her tail's wagging furiously. She's happier than ever. I smile through tears as I take comfort in knowing that she is set free to roam, and whenever I feel lost all I need to do is close my eyes, take a breath, and find her at home within my heart, along with all those who I've loved and lost.

"Daddy, I'm always here, right in your heart," she tells me.

And then other voices join in.

"We all are."

Epilogue

A year later, and Hood River is beginning to feel a bit like home. On a clear day, I see snowcapped Mount Adams to the north and the magnificent Mount Hood to the south. Wind rips through the gorge and creates conditions perfect for windsurfers and kiteboarders on the Columbia River. The entire community is full of outdoorsy folks and dog lovers. Today I'm at a coffee shop, writing. My phone vibrates with an incoming text message:

> Kristen: I don't feel like cooking tonight. Would you mind picking up a pizza if I order one?
> Rob: Pizza sounds delicious. Let me know when you put in the order, and I'll start packing up.

Earlier this morning I walked dogs at the shelter, something Kristen helped me work up the courage to do a few months back. I'd wanted to for a while, but I was afraid my heart couldn't take it. These days I look forward to spending time with the rescue dogs, and the volunteers at the dog shelter are becoming like family to me.

Most of my time nowadays is spent writing and taking pictures, trying to slow things down and stop the outward search, remembering the peace I found on those days with Bella when we simply enjoyed being alive. I usually find myself writing about my most recent experiences, and hope to use my writing and photography to make a difference. Today I'm giving a voice to the shelter dogs. When I look into the eyes of these dogs as they're caged behind metal fences, I can see Bella looking back at me. Sadly, some of these dogs get adopted for a short while then returned because they're "too much work." It breaks my heart as I imagine Bella being dropped off in a strange place with no explanation. So I write new stories in ways I hope encourage people looking to adopt a dog to make sure they find the right dog that fits their home, family, and lifestyle. They need to be willing to put in the work to give their new furry family member a fighting chance. Dogs are part of our world, but to the dogs, we are their entire world.

I'm learning, even now, that much of my unending search for purpose was really a search to *define* what purpose is. I've seen how one of the things I was searching for was staring me back in the face the whole time. The love Bella gave me, and the peace I felt when focusing on that love. That affected my attitude toward myself, the people I care about most — in fact, everyone and everything. I doubt my depression will ever disappear completely, but at least I can now identify it and understand it — and my wounds certainly don't need to dictate and limit the rest of my life. Under the big skies of the open road, Bella and I seemed to live as free as anyone has ever lived. Perhaps that freedom allowed us to learn to live outside the confines of our fears and live instead under the big open space of possibility.

Kristen: Pizza ordered. Ready in 15.
Rob: Groovy. Packing up.

I hit "post" on what I've just written about the shelter dogs, close my laptop, and pack it into my coyote-brown backpack. The same backpack I carried on my entire journey across the country with Bella. The Pacific Northwest rain pours heavily as I exit the coffee shop and run over to Ruthie with the hood of my jacket pulled over my head.

Ruthie's stickers remind me of our incredible journey. Asheville, Lake Placid, Tetons, Moab, Grand Canyon, Badlands, Glacier National Park, Death Valley. Mile marker zero from the Keys and Highway 101 from the Olympic Peninsula are my two favorite stickers, a true testament of the cross-country journey. I can only imagine how many adventures are yet to come.

I drive down the steep hill toward the pizza shop. My stomach is growling and I'm ready to eat, but as I pull into a parking spot, I notice a large yellow Lab wandering under the awnings of the strip mall, probably trying to stay dry. The dog has a collar on, but he's all alone. Traffic is brisk in this area of town, and I fear he'll wander out into the intersection and get hit.

I throw Ruthie into park, hop out, and give a quick whistle. The lab stops walking, stares at me a moment, then starts heading toward me. When I start walking toward him he hesitates, then starts heading in my direction again. Maybe I should just leave him alone.

I kneel on my haunches and examine the Lab. He lets me scratch his noggin and pet his coat. The rain has soaked through his fur, and his coat is cold. His collar has no tags. He's been out in the rain for a while. I glance at the two streets that form the corner of the parking lot. Cars whiz by in the rainy evening. I can't leave him here. I just . . . can't.

"Well, bud, you wanna come hang out and dry off a bit?" I ask.

He comes right along as I gently lead him by the collar toward Ruthie. When I open the back door of my 4Runner, the seats are already folded down flat, and a blanket intertwined with Bella's brown hairs is laid out across them. The lab looks older than Bella, past twelve or thirteen years old, and he hops right in and curls up on Bella's blanket, content to be anywhere but roaming outside in the rain.

Now what?

Pizza!

"I'll be right back!" I say to the Lab. "Promise."

I give him a quick pet, then run into the pizza shop, pick up my order, and ask the dude behind the counter: "Anyone lose a big ol' yellow Lab that you know?"

"Nah," he says. "But there's a trailer park across the way where dogs get loose. Sometimes they roam over here."

I say thanks and bring the pizza back to the 4Runner. The old Lab is still lying in the same spot I left him. I drive across the street and pull into the trailer park. It's tucked behind a few businesses, and the driveway is dark and narrow. Three plots in sits a trailer with the lights on inside. I park and knock, hear a child's footsteps scamper across the linoleum. A middle-aged Hispanic man opens the door. He doesn't speak English, but he motions to the boy to translate for us.

"We don't have any dog here," the boy says. "But they do in that house over there." He points across the street.

"Gracias para ayuda me. Buenos noches," I say in my attempt to speak the little Spanish I've picked up over the years, and head across the street. My knock sets off a series of loud barks. They have a dog, all right. A young boy opens the door only far enough to peek outside.

"Hi, bud! Do you guys have another dog? A yellow Lab?"

The boy solemnly shakes his head.

"Anyone else in the neighborhood have a dog?"

Another shake. Another no.

I walk back to the 4Runner and peer into the back window. The old fella is still resting in the back. I hop in the driver's seat and see that the pizza has remained intact.

"Thanks, buddy," I say. "I'd keep looking, but I've gotta get this pizza home before it gets cold. If I don't, I'll be out on the street tonight with you!"

Kristen and I are fostering a different dog in her apartment. He's a lovable snuggle bug but not too keen on traveling or meeting new people, so we're providing him with a home until he can be adopted by someone content with staying at home, who doesn't have many visitors.

"Here's the pi-zz-aa!" I say as I come inside Kristen's apartment. I say it with lots of enthusiasm, hoping a little charm will help along the way, and add, "Oh — and I just happen to have a giant yellow Lab in the 4Runner too."

Kristen looks at me quizzically. She's learned already that these surprises are all part of the fun. "Well — bring him inside," she says.

"You're not mad?"

"Of course not. I've probably done the same thing a dozen times myself." She walks to the closet in one brisk motion, tosses an old towel to me, and gets out a set of old sheets to protect the couch.

I stop and stare, awestruck that this exchange about bringing home a big, strange, wet dog is going so smoothly.

We put the harness on our foster pup so we can introduce the two dogs slowly, and I find a spare leash for the Lab, then head back to Ruthie. The old Lab follows me to the yard without hesitation. Our rescue pup is already there, and the two dogs sniff each other, and our rescue pup initiates play, but the old Lab looks like his playing days are behind him. We open the door to our cozy little home, and both dogs walk inside together, the old man sniffing his new surroundings, the younger pup fixated on sniffing his new guest.

We dry off the Lab and lay out a bed for him. He sniffs near the food bowl, so we offer him some kibble. He gobbles it up, then moseys over to the bed and plops down with his front paws crossed. He knows the score. It's clear he has a home somewhere.

We take a few good photos and send them to a manager at our shelter. She posts the picture on their website. "Found Dog!" Our next plan is to call the sheriff, who'll come and get the Lab. But the sheriff will need to leave him in a kennel overnight until volunteers can process the dog the next morning at eight o'clock. So we decide we don't want to do this to the Lab. He's getting along well enough with our foster pup. We can keep him, at least for one night.

By now our pizza is cold. I turn on the broiler and toss it in the oven. In a few minutes, it's back to piping hot. Kristen and I sit on the couch, stuffing our faces.

"What would you be doing right now if Franklin had gone missing?" I ask between bites.

"Same thing you'd be doing if Bella had been missing," she says.

We both set down our pizza. What if the Lab's owner hasn't thought about the shelter? What if he's still out searching in the rain? What if he doesn't use Facebook and can't see the picture of his pup being shared among the three thousand people who follow the shelter's page? What if he doesn't have enough money for the sheriff's fee to regain custody tomorrow?

I look at the old Lab. Though I'm sure he appreciates some food in his belly and a place to dry off and warm up a bit, he looks to be growing nervous, probably wondering if he's ever going to see his family again.

He hears my keys jingle, sees me putting on my coat. His ears perk up. I snap the leash back on his collar, and his tail wags with approval.

"All right, buddy," I say. "Let's get back out there and see if we can't find where you belong!"

"Good luck!" says Kristen to me. She gives me a kiss on the cheek while I head out the door.

It's still raining. Nearly dark now.

As I start driving, I notice the Lab's snout is up by the window, so I roll the rear window down, and he hangs his head outside. My heart smiles as I see a familiar scene. The Lab smiles as his ears flap in the wind.

We pull into the same parking lot, and I leave the Lab in the back while I go into each business in the strip mall and ask around but discover nothing. I return to Ruthie and open the back door. The Lab hops out and starts walking on leash with me. I do my best not to lead him in any specific direction to see if he'll lead me where he needs to go.

Sure enough, he leads me toward the busy street, the street I feared he would try to cross earlier. But right before we reach the intersection, a young man comes running across the street. "Manny! Manny!"

The old Lab's head perks up, and his tail begins to swing back and forth harder than any food or warm place have caused it to wag. So that's his name.

The young man is in his late twenties, early thirties. Jeans. Bushy beard. Flannel shirt. I wave him over.

"Oh my God," he says. "Where'd you find him?"

"He was right over here in the parking lot," I say. "You live nearby?"

"Two blocks away."

"I found him about an hour ago and didn't want to leave him alone on this busy street. I tried the neighborhood with no luck, so I took him home, dried him off, and fed him dinner. We posted his pictures online and were going to keep him overnight. Then I decided to come back here to see if he'd show me where he needs to be."

The man stares at me a moment. His eyes are wet, and he opens his arms for a hug. He draws me in, squeezes me tight. "My name's Alex," he says. "Thank you so, so much."

I hug him back, then we both let go. "Of course, man," I say. "I loved a dog with all my heart once. I get it."

He explains that his phone number is written on the inside of Manny's collar, a place I didn't think to check but will in the future. I encourage him to get tags for Manny, then we say our goodbyes and I watch as they wait for a break in traffic, then run across the street together. Alex is still wiping tears from his eyes. Manny is walking tall, head and tail both held high.

Then I walk back to Ruthie and send the Girl from Hood River a simple text:

Rob: He found his home.

ACKNOWLEDGMENTS

Wow. We made it. What an incredible journey this has been. Not only the journey that you've just finished reading, but the journey that was writing it all down. For nearly two years, I reached to the deepest parts of myself and put them into words the best I could. Since I had shared so much of Bella's and my journey as it happened, I wanted this book to explain the *why* behind it all. I wanted to share the stories that changed my perspective, stories that changed my life course, stories of Life, Love, and Loss. Some brought smiles as I wrote them, while others brought tears and heartache. Bringing them to the page was cathartic, but shaping them into a book was a daunting task, which I needed help with. I was blessed with the opportunity to work with an incredible collaborator and mentor, Marcus Brotherton, who drew out the stories that seemed the most meaningful, narrowed them down, and put them in an order that could flow into a single manuscript. Even then, many stories fell to the editing room floor, but that doesn't mean they didn't happen, or that they weren't meaningful.

To my friends and family. Nothing in my life would be possible without your unwavering support and encouragement. I am truly blessed in having so many wonderful people in my corner to pull me up when I am down, and to keep me grounded if I fly too far into the clouds. I am constantly amazed at how many of you stick with me through all of the ups and downs of this crazy rollercoaster called life, and I am forever grateful for your kindness, wisdom, comradery, and compassion. Thank you, for helping mold me into the man I've become, and continually strive to be. I've got a lot of work to do, but knowing that you are there for me when needed makes that task feel a lot less daunting.

For those of you that I was able to include

on the page, I hope I did you justice. I wrote these stories from the memories of my own experience, from my own perspective, and I took the utmost care when writing about others and wanted to represent you in a way you would be proud of. I hope that you are indeed proud to be a part of this amazing journey, and I want to thank you, so very much, for being a part of what makes this book so special to me. The book is titled *A Dog Named Beautiful,* but the story isn't complete without all of you. Again, thank you. Thank you for allowing me to share glimpses of your stories to help me tell my own.

Along with Bella, I included many friends, family members, mentors, loved ones, even strangers in this story, and I was crushed that I couldn't find a way to include everyone. I wanted to bring you onto the pages with me. I wanted to thank you by sharing the lessons I learned from you, the perspective I gained from our conversations, the motivation I gained from your support, and the passion that was fueled by yours. If this is you, you know who you are. You know your place in the story, and your place in my heart. I couldn't have done this without you, and in this moment, as you read this paragraph, you are with me. I'm not sure where life will be taking me, but I know that I'll keep exploring, growing my perspective, talking with strangers, while taking photos and writing along the way. So, whether it's on my own website/blog or on paper, I'm sure that I'll be able to share my appreciation in one form or another. Until then, thank you for being a part of my life, it truly is better because of you.

To Team Bella. So many people banded together from places far and wide through the online following during our trip. You came together to raise funds that allowed me to get the best treatments possible for Bella without worry of the costs, which is something I've tried to pay forward to others as my way of giving back. The gifts that you sent after Bella passed are treasures that I will cherish forever. During our journey, you invited Bella and me into your homes and as my own story continues, your doors and hearts have remained open. I hope to have the opportunity to accept those invitations and see and learn about your parts of the world. Thank you, for making me feel at home wherever I go.

To Bella, my beautiful baby girl. Just writing your name brings a smile to my face and warmth to my heart. You really were the most perfect being that I've ever encountered on this earth, you'll be a hard one to live up to. Thank you, for all of the love, the laughs, the adventures, and the forgiveness that you gave me and so many others over the years. You allowed me to learn to love and forgive myself, and for that I owe you everything. I also want to thank you for the opportunity to share so many other stories with the world. It was your smiling face and inspiring happy hop that drew people in from across the globe, and anyone who reads this book will likely read it because they wanted to learn more about your life. While learning about you, they learned about some of my best friends and family, and of course, Mike and Charity. Lives taken too soon that will now live on, because of you. Thank you, baby girl. Thank you for all that you've done and continue to do. Thank you for helping me find my heart, and my voice. Everyone could use a girl like you in their lives. Daddy loves you, and he's so damn proud. Don't forget to visit every now and again.

To Mike, my brother, mentor, and hero. Well, bro, I did it. I chased my dreams. They evolved along the way, sometimes in ways I didn't quite understand, but I never stopped following them. I refused to listen to the critics who didn't understand the mission that I set out on after reading your words. This mission has taken me to so many different places and connected me with so many amazing people that I would have never been able to experience without you, without your sacrifice. I truly hope that I've made you proud down here on this blue and green speck, and there isn't a day that goes by that I don't wish you could be alongside me on these adventures. Yet I take solace in knowing that you are with me wherever I go. I tell your story often, so that others know why I am who I am and do what I do, and now I've been able to share it in a way that will allow you to live on in the hearts of many, forever. The night Bella passed, I know that wasn't a dream, I truly believe it was a visit, and it gave me faith in an existence after this life. Thank you so much for taking care of Bella for me until I get there, I know she's in great hands. Rub her noggin and scratch her rump for me, would ya? I love you, brother, and can't wait until the moment I see you again. Until Valhalla.

To Charli. Thank you for everything you and your family have done for me; you all showed me what family can be. Thank you for supporting and encouraging Bella and me to take our trip. She was a perfect dog and you had such a big role in that. Again, thank you for allowing me to spend so much time with her. I'll never forget that kindness.

To Kristen. Though "The Girl from Hood River" has a fun and mysterious vibe, you are so much more than a beautiful part of my story. You are one of the strongest, most capable women I've ever met, and those who call you family and friend are all fortunate to have you in their stories. I'm in awe of all that you've done and am excited to see what is to come. You have no limits.

To the military and veteran community. It's obvious that, when looking back on my service, I struggle with what I didn't do, rather than being proud of what I did do. Most of this stems from survivor's guilt from my brother's death. It's hard not to let that one moment shape the memories and entire perspective when looking back. However, when clearing the cobwebs of the past and looking around me today, I see all of you. I see how so many of you are finding ways to continue to serve others. I see you taking your experiences and making the world a better place by sharing your perspectives to help others expand their own. I see you extending your hands to pull someone back to their feet. I've been that someone. I see you opening the doors to your homes, doors I've walked through and had more raw and honest conversations at your dinner tables than I have anywhere else in life. When I see all of you, and what you do for your communities, for your country, for the world, I am proud. I am proud to be amongst you. I'm proud to know that you'll always have my back. Know that I will always have yours in return. Thank you. Thanks for your service, and thank you for your friendship. This trip wouldn't have been possible without your support.

To Marcus Brotherton, James Melia of Flatiron Books, Andrea Henry of Penguin Publishing, and Cait Hoyt of Creative Artists Agency, thank you for championing this project and giving me the opportunity to put these stories on paper. I know I wasn't easy to work with, as I had so much I wanted to share, but you stuck with me until the end and together we were able to create something beautiful. Thank you so much for this opportunity to share not only Bella, but the biggest pieces of myself with so many. I am honored to have worked alongside you and am eternally thankful for all you have done to see this through from start to finish.

To everyone who has read this book, as you are holding this in your hands and reading these words, you are now a part of the story. You've taken this journey with us. You're connected with Bella, with me, with all of the people in the story, and now with each other. That's just one of the many beautiful aspects of the love of a dog, it connects us as people. It's a love that knows no boundaries and is translatable to every language on earth. It's a love that can teach us so much about what matters in this life we've been given. Thank you. Thank you for taking the time to read these words, and for learning the *why* behind our journey. I can only hope that sharing this may help others connect with each other and share their own stories of Life, Love, and Loss. It is those stories that wash away prejudice and allow us to see the more connected we are, the more we treat each other like family, and the more this entire planet feels like home. I know I'm an idealist, but I can't deny what I believe, and if I have this one chance to share it, I'd be a fool not to. So again, I thank you for taking the time to read all of the stories that I shared in this book, and if you would like to continue to follow wherever the journey may go after this, or simply see more of the photos of Bella and scenery from our trip, you can stay connected through my website:

rklifeillustrated.com.

Last, I want to list a few of the organizations and businesses that supported us in one way or another. Again, we wouldn't have been able to do this without you, and I want to show my appreciation by including you here:

Team Rubicon - Project Hero - Team Red White and Blue - Exploring Roots - Marine Corps League, Cornhusker Detachment - American Legion, Hollywood Post 43 - Veterans in Media and Entertainment - Adaptive Adventures - Wilderness on Wheels - Four Seasons Veterinary Specialist - Colorado State University Veterinary Teaching Hospital - Promise 4 Paws - Stryder Cancer Foundation - Baunj Strong Foundation - Underdog Alliance - "The Roles of Dogs in Our Society" — Amanda Hoeneman - Outward Bound - Valhalla Dive Group - Team Overland - CVT Tents - Ruff Wear - BOTE Boards — Toyota of Fort Walton Beach

ABOUT THE AUTHOR

Medically-retired US Marine Staff Sergeant **Rob Kugler** is a storyteller, photographer and writer. When he's not on the road, he lives in Nebraska.

The employees of Thorndike Press hope you have enjoyed this Large Print book. All our Thorndike, Wheeler, and Kennebec Large Print titles are designed for easy reading, and all our books are made to last. Other Thorndike Press Large Print books are available at your library, through selected bookstores, or directly from us.

For information about titles, please call:
(800) 223-1244

or visit our website at:
gale.com/thorndike

To share your comments, please write:
Publisher
Thorndike Press
10 Water St., Suite 310
Waterville, ME 04901